Metamaterial Electromagnetic Wave Absorbers

Synthesis Lectures on Electromagnetics

Editor
Aklesh Lakhtakia, *The Pennsylvania State University*

Metamaterial Electromagnetic Wave Absorbers
Willie J. Padilla and Kebin Fan
2021

Theory of Graded-Bandgap Thin-Film Solar Cells
Faiz Ahmad, Aklesh Lakhtakia, and Peter B. Monk
2021

Spoof Plasmons
Tatjana Gric
2020

The Transfer-Matrix Method in Electromagnetics and Optics
Tom G. Mackay and Akhlesh Lakhtakia
2020

© Springer Nature Switzerland AG 2022

Reprint of original edition © Morgan & Claypool 2022

Metamaterial Electromagnetic Wave Absorbers

Willie J. Padilla and Kebin Fan

ISBN: 978-3-031-03755-9   paperback
ISBN: 978-3-031-03765-8   PDF
ISBN: 978-3-031-03775-7   hardcover

DOI 10.1007/978-3-031-03765-8

A Publication in the Springer series
SYNTHESIS LECTURES ON ELECTROMAGNETICS

Lecture #4
Series Editor: Aklesh Lakhtakia, *The Pennsylvania State University*
Series ISSN
Print 2691-5448   Electronic 2691-5456

Metamaterial Electromagnetic Wave Absorbers

Willie J. Padilla
Duke University

Kebin Fan
Nanjing University, China

SYNTHESIS LECTURES ON ELECTROMAGNETICS #4

ABSTRACT

Electromagnetic metamaterials are a family of shaped periodic materials which achieve extraordinary scattering properties that are difficult or impossible to achieve with naturally occurring materials. This book focuses on one such feature of electromagnetic metamaterials—the theory, properties, and applications of the absorption of electromagnetic radiation.

We have written this book for undergraduate and graduate students, researchers, and practitioners, covering the background and tools necessary to engage in the research and practice of metamaterial electromagnetic wave absorbers in various fundamental and applied settings. Given the growing impact of climate change, the call for innovations that can circumvent the use of conventional energy sources will be increasingly important. As we highlight in Chapter 6, the absorption of radiation with electromagnetic metamaterials has been used for energy harvesting and energy generation, and will help to reduce reliance on fossil fuels. Other applications ranging from biochemical sensing to imaging are also covered. We hope this book equips interested readers with the tools necessary to successfully engage in applied metamaterials research for clean, sustainable energy.

This book consists of six chapters. Chapter 1 provides an introduction and a brief history of electromagnetic wave absorbers; Chapter 2 focuses on several theories of perfect absorbers; Chapter 3 discusses the scattering properties achievable with metamaterial absorbers; Chapter 4 provides significant detail on the fabricational processes; Chapter 5 discusses examples of dynamical absorbers; and Chapter 6 highlights applications of metamaterial absorbers.

KEYWORDS

electromagnetic wave absorbers, metamaterial absorbers, dynamical absorbers, lithography

To the pursuit of truth and knowledge.

We are like dwarfs on the shoulders of giants, so that we can see more than they, and things at a greater distance, not by virtue of any sharpness of sight on our part, or any physical distinction, but because we are carried high and raised up by their giant size.

–Bernard of Chartres, 1115

Contents

Preface

The advances in electromagnetic metamaterials and metasurfaces in just the past 20 years has been extraordinary. The word metamaterial is now common parlance in many areas of scientific study and, due to the later development of acoustical metamaterials and mechanical metamaterials, the term *electromagnetic metamaterials* is often used in order to disambiguate these different areas of inquiry. In the past two decades, an enormous body of work has been published on electromagnetic metamaterial absorbers operating across a huge swath of the spectrum. Even from the time we started to write this compendium, significant results appeared which we felt were important to include, thereby further delaying completion. Although the field of metamaterial electromagnetic wave absorbers is still fast growing, we felt it was important to give a brief description of the theory, achievements, fabricational techniques, and applications as it stands today. Even with this limitation, the material covers six chapters, and there are a few topics that have been omitted. Nonetheless, we have endeavored to reference all relevant works on the topic, and we ask our colleagues and friends for their understanding and forgiveness for any oversight in this regard.

Modern electromagnetic metamaterials research is significantly more efficient, accurate, and routine than it was at the start of the 21st century. The advent of the computer—especially the personal computer in 1975—bolstered the successful implementation of many numerical methods which seek solutions of Maxwell's equations. Although an algorithm to solve for vector electromagnetic fields on rectilinear grids was published in 1966—the Yee algorithm [1]—it wasn't until 1988 that the finite-difference time-domain method was widely popularized by Taflove [2]. Computational electromagnetic (CEM) simulation has now become routine with accurate and cost-effective options widely available using increasingly sophisticated techniques. During the same period, nanolithography continued to mature—as well as several new techniques becoming available—and metamaterials research has certainly benefited. The use of mercury lines for optical lithography in the 1980's permitted increasingly finer resolution (g-line $\lambda = 436$ nm, h-line $\lambda = 405$ nm, and i-line $\lambda = 365$ nm), and in the 90's the switch to deep ultraviolet excimer lasers realized 284 nm (KrF), and 193 nm (ArF) wavelength capability. [3] These advances were particularly important for pushing metamaterials to ever smaller-length scales and shorter wavelength operation. Nowadays, once conceived, an electromagnetic metamaterial can be simulated, fabricated, and characterized within only a few days at microwave frequencies. Since metamaterials may be scaled to operate over much of the electromagnetic spectrum, these initial microwave designs may be relatively easily modified to operate in other bands.

Finally, let us briefly mention the architecture and content of the book. Our primary objective is to provide an overview of the theory, simulation, experimental verification, and application of metamaterials and metasurfaces as absorbers of electromagnetic radiation. In the following chapters we present the Theory of Perfect Absorbers (Chapter 2) described by a number of approaches including: temporal coupled mode theory, surface current densities, multipole expansion, and interference theory. Chapter 3 introduces absorbers based on metallic metamaterials and all-dielectric based absorbers, followed by a description of their electromagnetic properties including: bandwidth, polarization dependence, and angular dependence. There are a number of available fabricational approaches used to fashion metamaterial absorbers—based on desired operational range of the electromagnetic spectrum—and in Chapter 4 we overview methods used including both masked and maskless lithography techniques. Metamaterials can be made to achieve their properties dynamically, and Chapter 5 discusses approaches to achieve tunability through electrical, mechanical, optical, and thermal control. Applications of metamaterial absorbers are discussed in Chapter 6, including various types of sensing, energy harvesting, and imaging. Although the pace of research on electromagnetic wave absorbers is rapid, we hope this book gives an accurate and timely survey of the field.

REFERENCES

[1] Kane Yee. Numerical solution of initial boundary value problems involving maxwell's equations in isotropic media. *IEEE Transactions on Antennas and Propagation*, 14(3):302–307, May 1966. DOI: 10.1109/TAP.1966.1138693 xiii

[2] Allen Taflove. Review of the formulation and applications of the finite-difference time-domain method for numerical modeling of electromagnetic wave interactions with arbitrary structures. *Wave Motion*, 10(6):547–582, December 1988. DOI: 10.1016/0165-2125(88)90012-1 xiii

[3] Stefan Landis, Editor. *Nano-Lithography*. John Wiley & Sons, Inc., February 2013. DOI: 10.1002/9781118622582 xiii

Willie J. Padilla and Kebin Fan
November 2021

Symbols

Symbol	SI Units	Description
d	m	Absorber thickness
$e = 1.602 \times 10^{-19}$	C	Charge of an electron
$c = 2.99792 \times 10^8$	m·s^{-1}	Speed of light in vacuum
f	s^{-1}	Frequency
$k_0 = \omega/c$	m^{-1}	Wavenumber of free space
$\epsilon_0 \approx 8.85 \times 10^{-12}$	F·m^{-1}	Permittivity of free space
$\tilde{\epsilon} = \epsilon_0 \tilde{\epsilon}_r$	F·m^{-1}	Complex permittivity
$\tilde{\epsilon}_r$	–	Complex relative permittivity
ϵ_∞	–	Epsilon infinity
γ_i	s^{-1}	Radiative loss rate for mode i
γ_e	s^{-1}	Scattering frequency of Lorentz ϵ oscillator
γ_m	s^{-1}	Scattering frequency of Lorentz μ oscillator
λ	m^{-1}	Wavelength
$\mu = 4\pi \times 10^{-7}$	H·m^{-1}	Permeability of free space
$\tilde{\mu} = \mu_0 \tilde{\mu}_r$	H·m^{-1}	Complex permeability
$\tilde{\mu}_r$	–	Complex relative permeability
μ_∞	–	Mu infinity
$\omega = 2\pi f$	rad/s	Angular frequency
$\omega_{0,e}$	rad/s	Center frequency of Lorentz ϵ oscillator
$\omega_{0,m}$	rad/s	Center frequency of Lorentz μ oscillator
ω_i	rad/s	Resonant frequency for mode i
ω_p	rad/s	Plasma frequency
δ_i	s^{-1}	Material loss rate for mode i
ρ	C·m^{-3}	Electric charge density
ς	m^{-2}	Density of multipoles per unit-cell area
$\tilde{\sigma}$	S·m^{-1}	Complex conductivity
σ_{sca}	m^2	Scattering cross section of a multipole
$\tilde{\chi}_e$	–	Electric susceptibility
$Z_0 = \sqrt{\mu_0/\epsilon_0} \approx 377$	Ω	Impedance of free space
$Z = \sqrt{\mu/\epsilon}$	Ω	Impedance
$Z_r = \sqrt{\mu_r/\epsilon_r}$	–	Relative impedance
$n = \sqrt{\epsilon_r \mu_r}$	–	Index of refraction
$\tilde{r} = |r|e^{i\theta_r}$	–	Reflection coefficient
$R = |r|^2$	–	Reflectance

$\tilde{t} = \|t\|e^{i\theta_t}$	–	Transmission coefficient
$T = \|t\|^2$	–	Transmittance
A	–	Absorptance
E	–	Emittance
$\mathbf{s}_{\text{in}}$	–	Wave inputs at the ports
$\mathbf{s}_{\text{out}}$	–	Wave outputs at the ports
N	m^{-3}	Electric carrier density
S	m^2	Unit-cell area
V	m^3	Volume per unit cell
$\mathbf{m}$	A·m^2	Magnetic dipole moment
$\mathbf{n}$	–	Unit vector along the radius vector $\mathbf{r}$
$\mathbf{p}$	C·m	Electric dipole moment
$\mathbf{r}$	m	Radius vector
$\mathbf{v}$	m·s^{-1}	Drift velocity of carriers
$\mathbf{B}$	W·m^{-2}	Magnetic flux density
$\mathbf{D}$	C·m^{-2}	Electric displacement
$\mathbf{E}$	V·m^{-1}	Electric field strength
$\mathbf{E}_i$	V·m^{-1}	Incident electric field strength
$\mathbf{H}$	A·m^{-1}	Magnetic field strength
$\mathbf{J}$	A·m^{-2}	Electric current density
$\mathbf{K_a}$	V/m	Asymmetric total electric surface current density
$\mathbf{K_e}$	A/m	Effective electric sheet current
$\mathbf{K_m}$	V/m	Effective magnetic sheet current
$\mathbf{K_s}$	V/m	Symmetric total electric surface current density
$\mathbf{P}$	C·m^{-2}	Polarization density
$\mathbf{T}$	C·m^2	Toroidal dipole moment
$\mathbf{a}$	–	Amplitude vector of a resonating system
$\mathbf{C}$	–	A direct non-resonant pathway from port to port
$\mathbf{D}$	–	Coupling matrix of output modes
$\mathbf{K}$	–	Coupling matrix of input ports and modes
$\widehat{M}$	A· m^3	Magnetic Quadrupole tensor
$\widehat{Q}$	A· m^{2}s^{-1}	Electric Quadrupole tensor
$\widehat{U}$	–	A 3 × 3 unit tensor
$\mathbf{S}_{ZRA}$	–	Scattering matrix for zero-rank absorbers
Γ	1/s	Total dissipation rate of the resonating modes
Γ_r	1/s	Decay rate of the resonating modes
Γ_i	1/s	Material loss rate of resonating modes
Ω	rad/s	A $n \times n$ matrix of ω_i's

CHAPTER 1

Introduction

1.1 MOTIVATION

This book deals with the absorption of electromagnetic radiation by metamaterials. Specifically, we are concerned with electromagnetic metamaterials, or metasurfaces, which are a type of composite material generally termed an artificial electromagnetic material, and their ability to achieved a designed absorption of electromagnetic waves. Several theories of metamaterial absorbers are presented in Chapter 2, while a brief introduction and history is given in Chapter 3. The more general history and theory underlying metamaterials is not covered, however we refer the interested reader to texts on the subject [1–3]. While this text broadly covers metamaterial absorbers operating across much of the electromagnetic spectrum, it is important to note that other common, (but more limiting), terms are often used to describe energy transfer (*heat transfer*), due to electromagnetic waves as *radiative heat transfer* and *thermal radiation* [4]. However, metamaterials possess a property termed *electromagnetic similitude* [5], thereby enabling them to control emission in other bands of the spectrum, not just in the infrared regime, and thus here we use the more general terms of *electromagnetic waves* or *electromagnetic radiation* to describe the radiative mechanism of energy transfer.

It is well known that energy (heat) may be transferred by three modes, i.e., that of conduction, convection, and radiation. However, convection is simply the bulk transport of a fluid (air), indicating that there are two fundamental energy transfer mechanisms—diffusion (the conduction of energy through molecular interactions) and radiation (the transfer of energy through photons or electromagnetic waves). The physics of conduction (diffusion) is a local phenomenon, whereas that of radiation is not, and may act over large distances. The physics describing the conversion of energy due to incident electromagnetic waves to heat, and their transport via diffusion and convection is only covered briefly. Rather, we focus on the electromagnetic properties (bandwidth, angle, polarization, etc.) achievable with metamaterial absorbers, and thus focus on free space electromagnetic waves which are incident on metamaterials, i.e., energy transport through electromagnetic radiation. We do, however, discuss a few applications of metamaterial absorbers where indeed the properties of convection and conduction are important, especially for energy harvesting and bolometric-based imaging.

Individual unit cells that constitute electromagnetic metamaterial are usually formed to be sub-wavelength, and therefore operate at wavelengths that are larger than their characteristic feature size, e.g., periodicity. As such, they are modeled as effective media [6], and described by an effective electrical permittivity ϵ_{eff}, and an effective magnetic permeability μ_{eff}. Due to their sub-wavelength nature, the electromagnetic properties of metamaterials are well described by the three fundamental radiative properties of *reflectance (R)*, *transmittance (T)*, and *absorptance (A)*. That is, although the diffuse scattering from metamaterials is small, but non-zero, use of the ending "-*ance*" is recommended by the National Institute of Standards and Technology (NIST) for rough surfaces, with the "-*ivity*" ending being reserved for perfectly smooth materials. Electromagnetic radiation incident upon a metamaterial can either be reflected, transmitted, or absorbed, and conservation of energy requires $R + T + A = 1$. All materials also emit electromagnetic radiation and—at a given temperature—a *blackbody* gives the maximum that can be emitted. The *emittance (E)* is given as the ratio of the energy emitted by a surface to that emitted by a blackbody, and thus E varies between zero and unity. The four radiative properties A, R, T, E may depend on frequency, temperature, angle from the surface normal, or polarization, and our goal in this book is to demonstrate that metamaterials do indeed possess the ability to tailor many of the fundamental absorptive and radiative properties of bodies and surfaces.

1.2   A BRIEF HISTORY OF ELECTROMAGNETIC WAVE ABSORBERS

Interest in absorbers of electromagnetic radiation largely began around the second half of the last century during World War II (WWII), and their importance in this connection was considerable. The high attenuation of radar possible with such absorbers enabled military advantages and consequently great technical progress was made. The role played by the materials themselves was primarily utilitarian, and little attention was paid to the scaling or universality of absorbers for their intrinsic interest. For example, radar was successfully used to find U-boats leading to losses. This in turn led to the development of radar absorbing materials (RAM) as an attempt to hide U-boat conning towers and periscopes. Two types of RAM were used—one termed a Wesch material, consisting of rubber sheets with carbonyl iron powder (commercially available before WWII), and the other a Jaumann absorber. Research into RAM was also undertaken as a means to increase the performance of radar systems. That is, it was applied to reflecting and scattering surfaces near the radar system, thereby reducing system noise. However, the approach used here was focused on anti-radar paints—called Halpern anti-radar paint (HARP) [7] which also used iron particles—as well as Salisbury absorbers [8].

Jaumann and Salisbury RAMs can be described as interference absorbers (Section 2.5), and both operate through use of a resistive screen (or screens in the case of the Jaumann absorber [9]) placed on a low loss and low dielectric constant ($\epsilon_r = 1.03 - 1.1$) spacer a distance $d = \lambda_0/4$ above a ground plane, where λ_0 is the operational wavelength [10]. The idea is that a portion of an incoming wave is partially reflected by the front resistive screen, and the remainder is reflected by the ground plane. Since the ground plane is spaced a distance of $d = \lambda_0/4$, the two waves are now out-of-phase at the resistive screen, and thus interfere (cancel) each other providing absorption. Since the Jaumann absorber uses multiple resistive screens, it can achieve greater bandwidth compared to single resistive sheet designs. A different design known as the

Dällenbach absorber uses a homogeneous lossy dielectric above a ground plane. One attempts to use a spacer material that is impedance matched to free space, i.e., $\epsilon_r = \mu_r$, thereby enabling zero reflectance and providing *dielectric loading* which reduces the wavelength $\lambda = \lambda_0/n = \lambda_0/\sqrt{\epsilon_r}$, and therefore the required thickness d.

The Jaumann RAM operated in the microwave regime from 2–15 GHz, i.e., across the S (2–4 GHz), C (4–8 GHz), X (8–12 GHz), and K_u (12–18 GHz) bands. The Jaumann absorber was 7.6 cm thick and thus achieved a wavelength to thickness ratio of $\lambda_0/d = 0.26 - 2$. The Wesch absorber operated in the S-band near 3 GHz ($\lambda_0 = 10$ cm), and had a nominal thickness of $d = 7.6$ mm, where the front surface was crumpled to achieve greater bandwidth [11]. The Wesch material thus achieved a wavelength to thickness ratio of $\lambda_0/d = 13$ [12]. The HARP absorber is composed of conducting particles, which may include aluminum, copper, permalloy, and graphite flakes, all dispersed in an insulating matrix of materials including waxes, synthetic rubbers, and other polymers. One variant of HARP operated in the X-band and had a thickness of 1.8 mm and thus achieved $\lambda_0/d = 17$ at band center [12].

In the 1950s, research turned both to expanding the capability of interference absorbers, and development of absorbers for anechoic chambers. The realization of so-called circuit analog (CA) absorbers, also termed frequency selective surfaces (FSS) [13], replaced the resistive screens of interference absorbers, and both resistive and reactive periodic components were used [9, 14–16]. The CA absorbers usually consisted of periodic elements above a conductive ground plane, and were modeled as an equivalent circuit consisting of an RLC series combination [13]. Many various types of periodic element shapes were explored including: dipole elements, N-pole elements, loop types, plate types, folded dipole elements, crosses, Jerusalem crosses, and combinations thereof [13]. CA absorbers where also used in multi-layer designs, i.e., two or more FSS layers above a ground plane, similar to that of the Jaumann absorbers. Absorbers were also developed to achieve lower reflectivity coefficients for anechoic chambers, and well as for operation at lower frequencies [11, 12, 17, 18].

Absorbers were pushed to the lower frequencies of VHF (30–300 MHz) and UHF (300–3,000 MHz) bands in the 1960s. Use of the standard interference type absorber, based solely on dielectric materials, would require a thickness of around a meter. However, use of ferrite and other magnetic materials embedded within the absorber permitted a reduction in the thickness [19]. These magnetically loaded absorbers were researched for potential application to anechoic chambers operating at low frequencies, and consisted of sintered ferrite plates above a ground plane, triple-layer structures consisting of sintered ferrite, dielectric, and sintered ferrite, and rubber ferrite [20]. Research on incorporation of magnetic materials into absorbers continued into the 1980's and 1990's, and various materials were explored, including Co-Ti substituted Ba hexaferrite [21], mixtures of rubber, carbon, and ferrite [22, 23], and nickel-zinc ferrite [24].

1.3   REFERENCES

[1] Constantin Simovski and Sergei Tretyakov. *An Introduction to Metamaterials and Nanophotonics*. Cambridge University Press, September 2020. DOI: 10.1017/9781108610735 1

[2] Ricardo Marqus, Ferran Martn, and Mario Sorolla. *Metamaterials with Negative Parameters*. John Wiley & Sons, Inc., December 2007. DOI: 10.1002/9780470191736 1

[3] Nader Engheta and Richard W. Ziolkowski, editors. *Metamaterials*. John Wiley & Sons, Inc., May 2006. DOI: 10.1002/0471784192 1

[4] C. Balaji. *Essentials of Radiation Heat Transfer*. Springer International Publishing, 2021. DOI: 10.1007/978-3-030-62617-4 1

[5] Julius Adams Stratton. *Electromagnetic Theory*. John Wiley & Sons, Inc., October 2015. DOI: 10.1002/9781119134640 1

[6] Tuck C. Choy. *Effective Medium Theory*. Oxford University Press, December 2015. DOI: 10.1093/acprof:oso/9780198705093.001.0001 1

[7] Otto Halpern. Method and means for minimizing reflection of high-frequency radio waves, February 2, 1960. US Patent 2923934A. 2

[8] Winfield W. Salisbury. Absorbent body for electromagnetic waves, June 1952. US Patent 2599944A. 2

[9] Ben A. Munk, Jonothan B. Pryor, and Yeow Beng Gan. On absorbers. In *Electromagnetic Materials*, pages 163–169, World Scientific, December 2003. DOI: 10.1142/9789812704344_0031 2, 3

[10] Eugene F. Knott, John F. Shaeffer, and Michael T. Tuley. *Radar Cross Section*. SciTech Publishing, 2004. DOI: 10.1049/sbra026e 2

[11] George T. Ruck, Donald E. Barrick, and Stuart William. *Radar Cross Section Handbook*, 1st ed., Peninsula Publishing, 1970. DOI: 10.1007/978-1-4899-5324-7 3

[12] K. J. Vinoy and R. M. Jha. Trends in radar absorbing materials technology. *Sadhana*, 20(5):815–850, October 1995. DOI: 10.1007/bf02744411 3

[13] Ben A. Munk. *Frequency Selective Surfaces: Theory and Design*. John Wiley & Sons, Inc., April 2000. DOI: 10.1002/0471723770 3

[14] R. Mittra, C. H. Chan, and T. Cwik. Techniques for analyzing frequency selective surfaces-a review. *Proc. of the IEEE*, 76(12):1593–1615, December 1988. DOI: 10.1109/5.16352 3

[15] Benedikt A. Munk, Peter Munk, and Jonothan Pryor. On designing Jaumann and circuit analog absorbers (CA absorbers) for oblique angle of incidence. *IEEE Transactions on Antennas and Propagation*, 55(1):186–193, January 2007. DOI: 10.1109/tap.2006.888395 3

[16] Arezou Edalati and Kamal Sarabandi. Wideband, wide angle, polarization independent RCS reduction using nonabsorptive miniaturized-element frequency selective surfaces. *IEEE Transactions on Antennas and Propagation*, 62(2):747–754, February 2014. DOI: 10.1109/tap.2013.2291236 3

[17] W. Emerson. Electromagnetic wave absorbers and anechoic chambers through the years. *IEEE Transactions on Antennas and Propagation*, 21(4):484–490, July 1973. DOI: 10.1109/tap.1973.1140517 3

[18] H. Severin. Nonreflecting absorbers for microwave radiation. *IRE Transactions on Antennas and Propagation*, 4(3):385–392, July 1956. DOI: 10.1109/tap.1956.1144419 3

[19] Y. Naito and K. Suetake. Application of ferrite to electromagnetic wave absorber and its characteristics. *IEEE Transactions on Microwave Theory and Techniques*, 19(1):65–72, January 1971. DOI: 10.1109/tmtt.1971.1127446 3

[20] Yoshiyuki Naito, Tetsuya Mizumoto, Yuuichi Wakita, and Michiharu Takahashi. Widening the bandwidth of ferrite electromagnetic wave absorbers by attaching rubber ferrite. *Electronics and Communications in Japan (Part I: Communications)*, 77(6):76–86, June 1994. DOI: 10.1002/ecja.4410770608 3

[21] N. Dishovski, A. Petkov, I. Nedkov, and I. Razkazov. Hexaferrite contribution to microwave absorbers characteristics. *IEEE Transactions on Magnetics*, 30(2):969–971, March 1994. DOI: 10.1109/20.312461 3

[22] S. Mirtaheri, J. Yiu, H. Seki, T. Mizumoto, and Y. Naito. The characteristics of electromagnetic wave absorbers composed of rubber carbon and ferrite. *The Transactions of IEICE*, E72(12):1447–1452, 1989. 3

[23] S. Mirtaheri, T. Mizumoto, and Y. Naito. The electromagnetic and dispersion characteristics of materials composed of rubber carbon and ferrite. *The Transactions of IEICE*, E73(10):1746–1759, 1990. 3

[24] H. M. Musal and H. T. Hahn. Thin-layer electromagnetic absorber design. *IEEE Transactions on Magnetics*, 25(5):3851–3853, 1989. DOI: 10.1109/20.42454 3

CHAPTER 2

Theory of Perfect Absorbers

A number of theories have been put forth to describe the underlying physics of metamaterial electromagnetic wave absorbers. Here we consider (Section 2.2) temporal coupled mode theory, (Section 2.3) effective electric and magnetic current sheets, (Section 2.4) multipole expansion, and (Section 2.5) interference theory. Only a brief review on these approaches is provided, and the interested reader is referred to references given in each section for more detail. Discussion of the fundamental limits on absorption bandwidth and thickness for metal backed absorbers is included in Section 2.6. We begin by defining the radiative properties relevant to a discussion of metamaterial electromagnetic wave absorbers, and state their dependence on the material parameters.

2.1   INTRODUCTION

A discontinuity in material properties significantly modifies the propagation of waves supported by the media. Consider, for example, plane waves in a medium consisting of two joined infinite half-spaces with differing material properties. In such a composite medium, a superposition of plane waves can be fashioned to represent an incident wave consistent with reflection and transmission (refraction) at the interface between the two media. The incident wave is then one which arrives from infinite depth in one of the media. A combination of additional waves is also required in order to satisfy the boundary conditions at the interface. These additional waves are called the reflected and transmitted waves. The reflected and transmitted waves must satisfy the causality law, and are thus unique. That is, although the plane waves considered in our example are steady-state, and thus extend throughout both media, the incident wave is taken to be the cause, and the reflected and transmitted waves are effects. Thus, causality requires that both the reflected and transmitted waves propagate away from the interface.

In Fig. 2.1a we show a plane interface between two different infinite half spaces, where Medium 1 is on the left (green), and Medium 2 is on the right (blue). Medium 1 and Medium 2 are described by complex material constants of $\tilde{\epsilon}^{(1)}$ and $\tilde{\mu}^{(1)}$, and $\tilde{\epsilon}^{(2)}$ and $\tilde{\mu}^{(2)}$, respectively. A plane wave is incident upon the interface from the left, and gives rise to reflected (left) and transmitted (right) waves. The form of these three waves is described by

$$\mathbf{E}_i(\mathbf{r}, t) = E_0 e^{i(\mathbf{k}_i \cdot \mathbf{r} - \omega t)}, \tag{2.1}$$

$$\mathbf{E}_r(\mathbf{r}, t) = E_{0,r} e^{i(\mathbf{k}_r \cdot \mathbf{r} - \omega t)}, \tag{2.2}$$

$$\mathbf{E}_t(\mathbf{r}, t) = E_{0,t} e^{i(\mathbf{k}_t \cdot \mathbf{r} - \omega t)}, \tag{2.3}$$

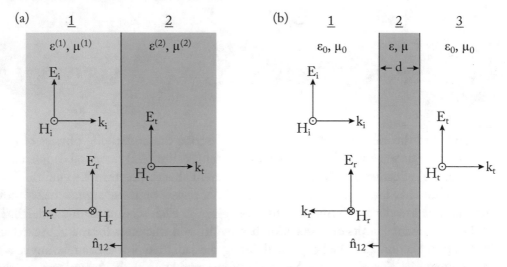

Figure 2.1: (a) Schematic of a plane interface between two infinite half spaces of different homogeneous media described by $\epsilon^{(1)}$ and $\mu^{(1)}$ on the left, and $\epsilon^{(2)}$ and $\mu^{(2)}$ on the right. (b) Slab of material ϵ and μ of thickness d embedded in vacuum.

where $\mathbf{E}_i$, $\mathbf{E}_r$, and $\mathbf{E}_t$ are the electric field strengths (V/m) of the incident, reflected, and transmitted waves, respectively, $\mathbf{r}$ is the position vector, $\mathbf{k}$ is the wavevector, ω is the angular frequency, and t is time.

The plane of incidence (POI) is defined as the plane which contains $\mathbf{k}_i$, $\mathbf{k}_r$, and $\mathbf{k}_t$. Due to the boundary conditions, the reflected and transmitted waves depend on the particular polarization of the incident wave. The polarization of these waves is defined with respect to the plane of incidence as transverse electric (TE) when the electric field is perpendicular (transverse) to the POI (also called S polarized), and transverse magnetic (TM) when the magnetic field is transverse to the POI (also called P polarized). The incident and reflected angles are defined with respect to the surface normal $\hat{n}_{12}$ of the interface, and in Fig. 2.1 normal incidence (zero degrees from the surface normal) is shown.

The behavior of electromagnetic fields in matter is determined by Maxwell's equations (MEs), and each ME gives rise to a boundary condition [1],

$$\nabla \times \mathbf{E} = -\frac{\partial \mathbf{B}}{\partial t} \quad \Rightarrow \quad \hat{n}_{12} \times (\mathbf{E}_1 - \mathbf{E}_2) = 0, \tag{2.4}$$

$$\nabla \times \mathbf{H} = \frac{\partial \mathbf{D}}{\partial t} + \mathbf{J} \quad \Rightarrow \quad \hat{n}_{12} \times (\mathbf{H}_1 - \mathbf{H}_2) = \mathbf{J}_s, \tag{2.5}$$

$$\nabla \cdot \mathbf{D} = \rho \quad \Rightarrow \quad \hat{n}_{12} \cdot (\mathbf{D}_1 - \mathbf{D}_2) = \rho_s, \tag{2.6}$$

$$\nabla \cdot \mathbf{B} = 0 \quad \Rightarrow \quad \hat{n}_{12} \cdot (\mathbf{B}_1 - \mathbf{B}_2) = 0, \tag{2.7}$$

where the left set are Maxwell's equations, and the boundary conditions are shown on the right. Here $\hat{n}_{12}$ is a surface normal which points from Medium 2 to Medium 1, and the bold characters denote time-varying vector fields and are real functions of spatial coordinates x, y, z (or equivalently the position vector $\mathbf{r}$), and time t. The subscripted 1's or 2's correspond to the fields in that medium, where $\mathbf{E}$ is the electric field strength (V/m), $\mathbf{B}$ is the magnetic flux density (V·s/m²), $\mathbf{J}$ is the electric current density (A/m²), and ρ is the electric charge density (A·s/m³). $\mathbf{D}$ and $\mathbf{H}$ are the auxiliary fields, and sometimes referred to as the electric displacement (A·s/m²), and the magnetic field strength (A/m), respectively. The subscripted "s" on $\mathbf{J}_s$ and ρ_s denote that it is only the surface terms that are important for the boundary conditions.

The microscopic polarization of a dielectric material is described by $\mathbf{P}(\mathbf{r}, t)$, and microscopic magnetization of a material by $\mathbf{M}(\mathbf{r}, t)$. The auxiliary fields are then defined by

$$\mathbf{D}(\mathbf{r}, t) = \epsilon_0 \mathbf{E}(\mathbf{r}, t) + \mathbf{P}(\mathbf{r}, t), \tag{2.8}$$

$$\mathbf{H}(\mathbf{r}, t) = \mu_0^{-1} \mathbf{B}(\mathbf{r}, t) - \mathbf{M}(\mathbf{r}, t). \tag{2.9}$$

Maxwell's equations Eqs. (2.4) and (2.7), and the auxiliary fields Eqs. (2.8) and (2.9) do not make any assumptions on the type of media, and are thus always valid. However, Eqs. (2.4)–(2.9) are not sufficient to determine the 12 components of $\mathbf{E}, \mathbf{B}, \mathbf{D}$, and $\mathbf{H}$. The so-called "constitutive equations" relate the auxiliary fields $\mathbf{D}$ and $\mathbf{H}$ to $\mathbf{E}$ and $\mathbf{B}$, and are often used to reduce the number of unknowns, however at the cost of assumptions about the media.

We may describe isotropic and linear media in the frequency domain where the polarization and magnetization take the form

$$\mathbf{P}(\mathbf{r}, \omega) = \epsilon_0 \chi_e(\mathbf{r}, \omega) \mathbf{E}(\mathbf{r}, \omega), \tag{2.10}$$

$$\mathbf{M}(\mathbf{r}, \omega) = \chi_m(\mathbf{r}, \omega) \mathbf{H}(\mathbf{r}, \omega), \tag{2.11}$$

here χ_e and χ_m are called the electric and magnetic susceptibility, respectively. Thus, dropping the explicit spatial and frequency dependence, we may write the constitutive equations as

$$\mathbf{D} = \epsilon_0 \mathbf{E} + \epsilon_0 \chi_e \mathbf{E} = \epsilon_0(1 + \chi_e)\mathbf{E} = \epsilon_0 \epsilon_r \mathbf{E} = \epsilon \mathbf{E}, \tag{2.12}$$

$$\mathbf{B} = \mu_0 \mathbf{H} + \mu_0 \mathbf{M} = \mu_0(1 + \chi_m)\mathbf{H} = \mu_0 \mu_r \mathbf{H} = \mu \mathbf{H}. \tag{2.13}$$

The permittivity is defined as $\epsilon = \epsilon_0 \epsilon_r$, and the permeability as $\mu = \mu_0 \mu_r$, where subscripted r's stand for relative values, $\epsilon_0 = 8.85 \times 10^{-12}$ (F/m), and $\mu_0 = 4\pi \times 10^{-7}$ (H/m).

Next, we consider an interface between two different media, as depicted in Fig. 2.1a, the definition of the reflection coefficient ($\tilde{r}$) is

$$\tilde{r} = \frac{\mathbf{E}_r}{\mathbf{E}_i} \equiv |r|e^{i\theta_r} \tag{2.14}$$

and the transmission coefficient ($\tilde{t}$) is

$$\tilde{t} = \frac{\mathbf{E}_t}{\mathbf{E}_i} \equiv |t|e^{i\theta_t}, \tag{2.15}$$

where θ_r and θ_t are the reflected and transmitted phases, respectively. The reflectance and transmittance are given as

$$R = |\tilde{r}|^2 \tag{2.16}$$

and

$$T = |\tilde{t}|^2 , \tag{2.17}$$

respectively. As mentioned previously, conservation of energy requires $A = 1 - R - T$, where A is the absorptance.

Figure 2.1b shows a homogeneous slab of matter thickness d embedded in vacuum with material parameters $\epsilon = \epsilon_0 \epsilon_r$ and $\mu = \mu_0 \mu_r$. Since there are now two interfaces, reflection and transmission may occur at each boundary. Nonetheless, the reflection and transmission coefficients are easily determined using the transfer matrix method, and at normal incident they are [2],

$$\tilde{r} = \frac{\frac{i}{2}\left[Z_r^{-1} - Z_r\right]\sin(nk_0 d)}{\cos(nk_0 d) - \frac{i}{2}\left[Z_r^{-1} + Z_r\right]\sin(nk_0 d)} \tag{2.18}$$

and

$$\tilde{t} = \frac{1}{\cos(nk_0 d) - \frac{i}{2}\left[Z_r^{-1} + Z_r\right]\sin(nk_0 d)} , \tag{2.19}$$

where $k_0 = \omega/c$ is the free space wavenumber, $Z_r = \sqrt{\mu_r/\epsilon_r}$ is the relative impedance, and $n = \sqrt{\epsilon_r \mu_r}$ is the index of refraction.

Equations (2.18) and (2.19) may be used to solve for the index of refraction and relative impedance,

$$n = \frac{1}{k_0 d}\arcsin\frac{1}{2\tilde{t}}\left[1 - \tilde{r}^2 + \tilde{t}^2\right] \tag{2.20}$$

$$Z_r = \pm\sqrt{\frac{(1 + \tilde{r})^2 - \tilde{t}^2}{(1 - \tilde{r})^2 - \tilde{t}^2}} . \tag{2.21}$$

The relative permeability and relative permittivity can then be determined from

$$\mu_r = n Z_r \tag{2.22}$$

and

$$\epsilon_r = \frac{n}{Z_r} . \tag{2.23}$$

A slab of material embedded in vacuum which realizes material parameters of $\epsilon_r = 1$ and $\mu_r = 1$ is termed *impedance matched*, i.e., $Z_r = 1$. In that case, from Eqs. (2.18) and (2.19) we have

$$\tilde{r} = 0 \tag{2.24}$$

and

$$\tilde{t} = e^{inkod} = e^{in_1 k_0 d - n_2 k_0 d}, \tag{2.25}$$

where we have expressed the index of refraction in real and imaginary parts using $\tilde{n} = n_1 + in_2$. We also consider a complex relative impedance given as $Z_r = Z_{1,r} + iZ_{2,r}$. Thus, for an impedance matched slab of material in vacuum the reflectance and transmittance are

$$R = |\tilde{r}|^2 = 0 \tag{2.26}$$

and

$$T = |\tilde{t}|^2 = e^{-2n_2 k_0 d} \tag{2.27}$$

and, therefore, the absorptance is

$$A = 1 - e^{-2n_2 k_0 d}, \tag{2.28}$$

which shows that if the material loss (n_2) is large, and the thickness d is sufficient, we may achieve $A \approx 1$.

In Fig. 2.2 we plot results of a slab of material thickness d which may be described by Lorentz oscillators in both ϵ_r and μ_r—with the frequency dependent form given by

$$\tilde{\epsilon}_r(\omega) = \epsilon_\infty + \frac{\omega_p^2}{\omega_{0,e}^2 - \omega^2 - i\gamma_e\omega} \tag{2.29}$$

and

$$\tilde{\mu}_r(\omega) = \mu_\infty + \frac{\omega_{p,m}^2}{\omega_{0,m}^2 - \omega^2 - i\gamma_m\omega}, \tag{2.30}$$

where ϵ_∞ is called "epsilon infinity" and μ_∞ is called "mu infinity" and both specify contributions to ϵ and μ, respectively, at frequencies higher than the range of consideration, $\omega_p = \sqrt{ne^2/\epsilon_0 m}$ is the plasma frequency, $\omega_{0,e}$ and $\omega_{0,m}$ are the center frequencies of the ϵ and μ oscillators, respectively, and γ_e and γ_m are the scattering frequencies of the ϵ and μ oscillators, respectively. Here, n is the number (carrier) density with units of [# carriers/m^3], $e = 1.6 \times 10^{-19}$ [Coulombs] is the charge of an electron, and m is the carrier mass—of order of an electron mass $m = 9.11 \times 10^{-31}$ kg.

In Fig. 2.2a–b we plot the real part of the relative impedance $Z_{1,r}$, and imaginary index of refraction n_2, both as a function of $\omega_{0,e}$ and $\omega_{0,m}$ at an observation frequency of $\omega = 2.5$ THz. As given by Eqs. (2.26) and (2.27), the reflectance is minimized when a material is impedance matched $Z_r = 1$, and the transmittance is minimized when n_2 is large. In Fig. 2.2a we observe that when the center frequencies of each oscillator are similar $\omega_{0,e} \approx \omega_{0,m}$, that the relative impedance is approximately equal to that of free space, i.e., the white contour line is plotted for $Z_{1,r} = 1$, and thus the reflectance is small. Likewise, in Fig. 2.2b we show that when $\omega_{0,e} \approx \omega_{o,m}$, n_2 is large, thereby minimizing transmittance. In Fig. 2.2 we have used the following

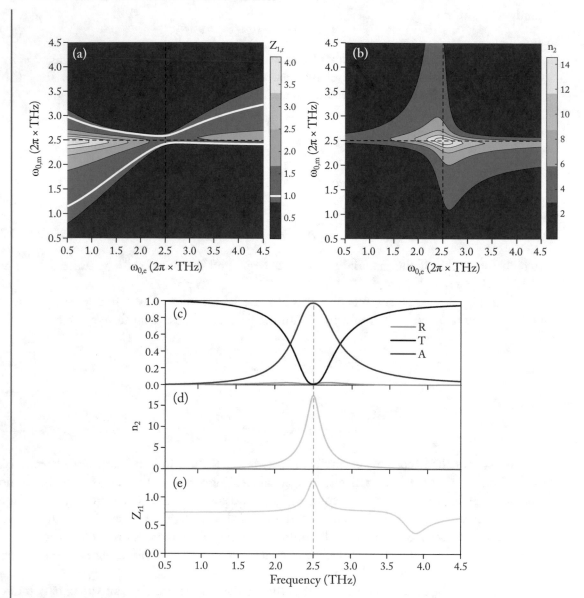

Figure 2.2: Dependence of absorbance for a slab of material described by Lorentz oscillators in ϵ_r and μ_r Eqs. (2.29) and (2.30), with parameters described in the text. (a) and (b) show the dependence of $Z_{1,r}$ and n_2 on $\omega_{0,e}$ and $\omega_{0,m}$, respectively. In (c) we show the frequency dependence of A, R, and T for parameters described in the text, and (d) and (e) show the corresponding n_2 and $Z_{1,r}$, respectively.

parameters for the Lorentz oscillators (see Eqs. (2.29) and (2.30)): $d = 2.5\,\mu\text{m}$, $\epsilon_\infty = 2.0$, $\omega_p = 4 \times 2\pi$ THz, $\omega - 2.5 \times 2\pi$ THz, $\gamma_e = 3.0 \times 2\pi$ TH, $\mu_\infty = 1.0$, $\omega_m = 3 \times 2\pi$ THz, $\gamma_m = 1.0 \times 2\pi$ THz. Next, we fix our center frequencies $\omega_{0,e} = \omega_{0,m} = 2.5 \times 2\pi$ THz, and plot the frequency dependence of the absorptance (red), transmittance (black), and reflectance (blue) in Fig. 2.2c. The transmittance and reflectance are both small at 2.5 THz, and thus the absorptance is peaked near unity. In Fig. 2.2d–e we plot the frequency dependence of n_2 and $Z_{1,r}$. We again find that at $\omega = 2.5 \times 2\pi$ THz n_2 is peaked leading to low transmission, and $Z_{1,r} \approx 1$, i.e., the material is approximately impedance matched.

2.2   TEMPORAL COUPLED-MODE THEORY

Temporal coupled-mode theory (TCMT) is a general framework that is useful for describing idealized systems, where a perturbation is added which represents a weak disturbance to the system. In TCMT for electromagnetic wave absorbers, the perturbation represents a weak coupling of the resonant systems together, and the method is analogous to time-dependent perturbation theory. The theory is thus general, and a variety of electromagnetic phenomena which originates from a coupling between resonators and, for example, waveguide ports can be addressed, including Fano resonances, and optical switching [3–5]. The particular form of TCMT used here describes a coupled resonating system with parameters representing resonant frequencies, and coupling constants without considering the details of the resonators and waveguides. In this section, we will briefly described TCMT utilized for perfect absorber design. More detail on TCMT applied to metamaterial absorbers is provided in Appendix A.

Since TCMT is a general framework it not only applies to a conventional two-port system—in which only one input and one output waveguide ports are considered—it may also be applied to a closed resonating system consisting of multiple ports and multiple resonant modes. To simplify our description of the perfect absorber without loss of generality, only two ports and a resonant cavity with two modes, are considered. Following the description of TCMT from Ref. [4], we begin by considering an incoming port wave which couples into a resonator and may exhibit a number of different loss mechanisms, including dissipative loss within the resonator, and port decay. Under these assumption, the dynamic equations for such a system, including ports, are given as:

$$\frac{da}{dt} = (-i\boldsymbol{\Omega} - \boldsymbol{\Gamma})\,a + \mathbf{K}^T s_{\text{in}}, \tag{2.31}$$

$$s_{\text{out}} = \mathbf{C}s_{\text{in}} + \mathbf{D}a, \tag{2.32}$$

where a is a vector with a time harmonic dependence of $\exp(-i\omega t)$, matrices $\boldsymbol{\Omega}$, and $\boldsymbol{\Gamma}$ are the resonant frequencies, and $\boldsymbol{\Gamma} = \boldsymbol{\Gamma}_r + \boldsymbol{\Gamma}_i$ is the total decay rate of the modes, with a radiative loss rate $\boldsymbol{\Gamma}_r$, and material loss rate $\boldsymbol{\Gamma}_i$. Matrices $\mathbf{K}$ and $\mathbf{D}$ correspond to input and output coupling of resonant modes to the ports, respectively. The vector s_{in} consists of elements which represent inputs from each port where $|s_{\text{in}}|^2$ is equal to the input power, with similar terms for the outputs

s_{out}. Matrix **C** represents a direct non-resonant pathway from port to port. From the TCMT equations, we may define a scattering matrix **S** as

$$S \equiv \frac{s_{\text{out}}}{s_{\text{in}}} = C + D[-i\omega I + i\Omega + \Gamma]^{-1}K^{T}, \tag{2.33}$$

where **I** is the identity matrix.

For a reciprocal cavity system with two physical ports, which is symmetric with respect to its mid-plane, both modes can be either even or odd. The scattering matrix is described as (see Appendix A),

$$S = C + DM^{-1}D^{T} = \begin{bmatrix} \tilde{r} & \tilde{t} \\ \tilde{t} & \tilde{r} \end{bmatrix} - \frac{1}{\det(M)} \left\{ P_2\gamma_1 \begin{bmatrix} 1 & e^{i\beta_1} \\ e^{i\beta_1} & 1 \end{bmatrix} \right.$$
$$\left. + P_1\gamma_2 \begin{bmatrix} 1 & e^{i\beta_2} \\ e^{i\beta_2} & 1 \end{bmatrix} \right\} \begin{bmatrix} \tilde{r} & \tilde{t} \\ \tilde{t} & \tilde{r} \end{bmatrix}, \tag{2.34}$$

where

$$M = \begin{bmatrix} L_1 & \gamma_0 \\ \gamma_0^* & L_2 \end{bmatrix} \tag{2.35a}$$

$$L_1 = -i(\omega - \omega_1) + \delta_1 + \gamma_1 \tag{2.35b}$$

$$L_2 = -i(\omega - \omega_2) + \delta_2 + \gamma_2 \tag{2.35c}$$

$$P_1 = -i(\omega - \omega_1) + \delta_1 + \frac{1 - e^{i\beta}}{2}\gamma_1 \tag{2.35d}$$

$$P_2 = -i(\omega - \omega_2) + \delta_2 + \frac{1 - e^{i\beta}}{2}\gamma_2 \tag{2.35e}$$

$$\beta = \beta_1 + \beta_2, \tag{2.35f}$$

and the nonresonant complex reflection and transmission coefficients are denoted by r and t. β_i, $(i = 1, 2)$ is the mode symmetry phase, i.e., for an even (odd) mode, $\beta_i = 0(\pi)$. The resonant frequencies are ω_1 and ω_2, and δ_1 and δ_2 are the intrinsic material loss rates. Decay rates of the modes to the ports are described as γ_1 and γ_2. For structures which break σ_h symmetry, the two modes are not orthogonal to each other anymore, resulting in a more complex response with extra off-diagonal entries in the matrix of Ω [6]. Additionally, if two modes are in the strong coupling regime, the off-diagonal entries of Ω are also non-zero. Tuning the coupling between two coupled resonators in a non-Hermitian system could break the PT-symmetry of the structure. Through further modifying the loss in the system, perfect absorption can also be reached in the broken PT-symmetry phase [7]. We next only discuss several metamaterial perfect absorber (MPA) cases, which are categorized based on the σ_h symmetry.

2.2.1 CASE I: ZERO RANK ABSORBER

It has been shown that perfect absorption cannot be achieved with only symmetric modes due to anti-crossing present in reciprocal systems [8]. Therefore, to achieve perfect absorption, two modes should be present, and possess opposite symmetries $\beta = \pi$ (even-odd or odd-even), i.e., no coupling between the two modes, $\gamma_0 = 0$. When the two modes are degenerate, that is $\omega_1 = \omega_2 = \omega_0$, the scattering matrix only depends on the decay rates and nonresonant transmission / reflection. With no loss of generality we assume mode 1 is even ($\beta_1 = 0$), and mode 2 is odd ($\beta_2 = \pi$). Equation (2.34) reduces to

$$
S = \left\{ I - \frac{\gamma_1}{-i(\omega - \omega_0) + \gamma_1 + \delta_1} \begin{bmatrix} 1 & 1 \\ 1 & 1 \end{bmatrix} \right. \\
\left. - \frac{\gamma_2}{-i(\omega - \omega_0) + \gamma_2 + \delta_2} \begin{bmatrix} 1 & -1 \\ -1 & 1 \end{bmatrix} \right\} \begin{bmatrix} \tilde{r} & \tilde{t} \\ \tilde{t} & \tilde{r} \end{bmatrix}.
\tag{2.36}
$$

When both modes are critically coupled, i.e., $\gamma_1 = \delta_1$ and $\gamma_2 = \delta_2$, at a resonant frequency of ω_0, Eq. (2.36) is

$$
S_{\text{ZRA}} = \begin{bmatrix} 0 & 0 \\ 0 & 0 \end{bmatrix}.
\tag{2.37}
$$

Thus, for a perfect absorber consisting of two degenerate critically coupled modes of opposite symmetry the nullity of the scattering matrix is two. To evaluate the total absorption in the system, we define the absorption matrix $A = I - S^+ S$, where at least one nonzero eigenvalue α exists for a lossy two port system. When $\alpha = 0$, no absorption occurs, but for $\alpha = 1$, the system perfectly dissipates incoming waves, i.e., perfect absorption. From Eq. (2.37), the absorption matrix is unitary, which corresponds to two absorption peaks, indicating that any linear super-postion of incoming waves from the two ports will be perfectly dissipated inside the structure.

2.2.2   CASE II: COHERENT PERFECT ABSORBER

If the resonator only supports a single critical-coupled mode at ω_0, e.g., an even mode, Eq. (2.34) can be further reduced with $\omega_2 \neq \omega_1 = \omega_0$, $\gamma_1 = \delta_1$, and $\beta = \pm\pi$, which gives a scattering matrix of

$$
\begin{aligned}
S_{\text{even}} &= \left\{ I - \frac{\gamma_1}{-i(\omega - \omega_0) + \gamma_1 + \delta_1} \begin{bmatrix} 1 & 1 \\ 1 & 1 \end{bmatrix} \right. \\
&\quad \left. \left. - \frac{\gamma_2}{-i(\omega - \omega_2) + \gamma_2 + \delta_2} \begin{bmatrix} 1 & -1 \\ -1 & 1 \end{bmatrix} \right\} \right|_{\omega=\omega_0} \begin{bmatrix} \tilde{r} & \tilde{t} \\ \tilde{t} & \tilde{r} \end{bmatrix} \\
&= (\tilde{r} - \tilde{t}) B \begin{bmatrix} 1 & -1 \\ -1 & 1 \end{bmatrix},
\end{aligned}
\tag{2.38}
$$

where

$$
B = \frac{1}{2} - \frac{\gamma_2}{-i(\omega_0 - \omega_2) + \gamma_2 + \delta_2}.
\tag{2.39}
$$

It is clear that **rank**$(S_{\text{even}}) = 1$. Similarly, we can also prove that **rank**$(S_{\text{odd}}) = 1$, i.e., the nullity of the scattering matrix for coherent perfect absorbers is 1. The absorption matrix for the single mode system is given as

$$\mathbf{A}_{\text{even}} = \mathbf{I} - \mathbf{S}_{\text{even}}^{+}\mathbf{S}_{\text{even}} = \mathbf{I} - 2|B|^2|\tilde{r} - \tilde{t}|^2 \begin{bmatrix} 1 & -1 \\ -1 & 1 \end{bmatrix}. \tag{2.40}$$

The eigenvalues of $\mathbf{A}_{\text{even}}$ are $\alpha_1 = 1$ and $\alpha_2 = 1 - 4|B|^2|\tilde{r} - \tilde{r}|^2 \neq 1$ since the reflection and transmission coefficients cannot be equal in both magnitude and phase. Thus, since only one eigenvalue of the absorption matrix $(\mathbf{A} - \mathbf{I})s_{\text{in}} = 0$ can be unity, this indicates that perfect absorption can only be achieved with input from two ports, where the inputs are equal in both magnitude and phase. With the same procedure, we can show that perfect absorption for an odd-mode resonator can be achieved if the input from two ports is equal in magnitude, but *out of phase*.

2.2.3   CASE III: GROUND PLANE ABSORBER

We consider TCMT for a metamaterial absorber which typically consists of a patterned planar metallic resonator space above a continuous conducting ground plane. The ground plane is approximated as a perfect electric conductor with negligible field penetration into the metal layer, and zero transmission, which is thus a single-port system. When the spacing between the resonator and the ground plane is optimal, the reflection from the planar resonator layer can be canceled through destructive interference with the reflection from the ground plane. Then Eq. (2.33) can be described with the coupling matrix $\mathbf{D} = \sqrt{2\gamma_1}e^{-i\theta_{11}}$ and $r = -e^{-i2\theta_{11}}$, where θ_{11} is the phase angle of matrix element D_{11}, which gives

$$\mathbf{S}_{GPA} = \tilde{r} - \frac{2\gamma_1}{-i(\omega - \omega_0) + \delta_1 + \gamma_1}\tilde{r} = \frac{-i(\omega - \omega_0) + \delta_1 - \gamma_1}{-i(\omega - \omega_0) + \delta_1 + \gamma_1}\tilde{r}. \tag{2.41}$$

It is clear that as long as the resonator is critically coupled with intrinsic loss rate balancing the radiation decay rate, the reflection coefficient becomes zero, indicating a perfect absorption state.

2.3   EFFECTIVE ELECTRIC AND MAGNETIC CURRENT SHEETS

A useful concept for describing the scattering of electromagnetic waves in planar metamaterials (metasurfaces) is that of electric and magnetic sheet currents, i.e., surface current densities, which describe both the conduction and polarization currents induced in matter via external $\mathbf{E}$ and $\mathbf{H}$ fields. Such a description using continuous current sheets was proposed by Wheeler in the late 1940's [9], and further developed in the 1960's [10] as a model for phased arrays. In 2003 the concept of capacitive loading between dipole elements of phased arrays was proposed, which had the effect of producing nearly continuous current sheets [11]. With regard to phased array performance, this may seem to be the opposite of what is desired, since traditionally researchers tried to minimize antenna coupling in the arrays. However, it should be noted that since antenna coupling is ultimately unavoidable, researchers began to take advantage of the coupling and introduced the concept of linear connected arrays, which resulted in approximately continuous current sheets with increased performance [12]. In 2003 it was shown that an impedance boundary condition can be used to describe planar arrays of small particles [13, 14], which has since been used to describe the condition of absorption in metamaterials and metasurfaces [6]. This approach to planar metasurfaces has been termed effective electric and magnetic current sheets, which we discuss next.

We begin from Ohms law given by

$$\mathbf{J} = Ne\mathbf{v} = \tilde{\sigma}\mathbf{E}, \tag{2.42}$$

where bold characters denote vectors, and $\mathbf{J}$ is the current density (a flux given by charge per unit area per unit time), $\tilde{\sigma}$ is the complex conductivity, and $\mathbf{v}$ is the drift velocity of carriers that constitute the current. Here N is the number (carrier) density with units of [# carriers/m^3], $e = 1.6 \times 10^{-19}$ [Coulombs] is the charge of an electron. The complex conductivity is related to the complex permittivity through $\tilde{\sigma} = i\omega(\epsilon_0 - \tilde{\epsilon})$, where $\tilde{\epsilon}$ is the complex permittivity defined by $\tilde{\epsilon} = \epsilon_0\epsilon_r = \epsilon_0(1 + \chi_e)$, where χ_e is the electric susceptibility, and $\epsilon_0 = 8.85 \times 10^{-12}$ F/m is the permittivity of free space. (Here we have used an $e^{-i\omega t}$ time-harmonic convention.) Substituting the permittivity and susceptibility into the conductivity in Ohms law we find

$$\mathbf{J} = i\omega\,(\epsilon_0 - \epsilon)\,\mathbf{E} = -i\omega\epsilon_0\chi_e\mathbf{E}. \tag{2.43}$$

The polarization density is $\mathbf{P} = d\mathbf{p}/dV$, where $\mathbf{p}$ is the dipole moment and V is the volume, is taken to be proportion to the electric field in the linear regime as

$$\mathbf{P} = \epsilon_0\chi_e\mathbf{E}. \tag{2.44}$$

Thus, the current density may be written as,

$$\mathbf{J} = -i\omega\mathbf{P}. \tag{2.45}$$

Finally, if we multiply Eq. (2.45) through by a unit of length, we have the effective surface averaged electric sheet current [6, 14]

$$\mathbf{K}_e = -i\omega\frac{\mathbf{p}}{S}, \tag{2.46}$$

where S is the unit-cell area. Likewise, we may derive a similar form for the effective surface averaged magnetic sheet current,

$$\mathbf{K}_m = -i\omega\mu_0\frac{\mathbf{m}}{S}, \tag{2.47}$$

where $\mathbf{m}$ is the magnetic dipole moment.

We consider a metamaterial consisting of two infinite parallel sheets, separated by a distance d, which supports $\mathbf{K}_e$ and $\mathbf{K}_m$. An electromagnetic wave is traveling in the $-\hat{z}$ direction

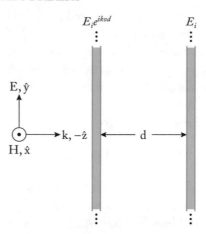

Figure 2.3: Schematic of a metamaterial absorber of thickness d, which is composed of two effective current sheets (blue) of infinitesimal thickness.

with its k vector parallel to the surface normal ($\hat{z}$) of the two sheets as depicted in Fig. 2.3. The first sheet which the wave arrives to gives rise to a surface current density $\mathbf{K}_{e,1}$ and $\mathbf{K}_{m,1}$, and $\mathbf{K}_{e,2}$ and $\mathbf{K}_{m,2}$ is induced on the second sheet. The second sheet is located at the origin, where the wave has an amplitude E_i. The phase delay between the two sheets, given by the distance d, gives rise to an amplitude of $E_i e^{ik_0 d}$, where k_0 is the free-space wave vector. The wave impedance is defined as $Z = E/H$, where E and H are the magnitudes of the $\mathbf{E}$ and $\mathbf{H}$ fields, respectively, and we define Z_0 to be the wave impedance of free-space, i.e., $Z_0 = \sqrt{\mu_0/\epsilon_0} \approx 377\Omega$. The magnetic field strength H and the surface current density K have the same units, thus we may write the electric field as $E = ZH = ZK_e$.

In order for a wave to be entirely absorbed, the total transmitted and reflected fields must sum to zero, given by

$$-\frac{Z_0}{2}\mathbf{K}_{e,1}e^{ik_0 d} - \frac{Z_0}{2}\mathbf{K}_{e,2} = -E_i \tag{2.48}$$

for the transmission, and

$$-\frac{Z_0}{2}\mathbf{K}_{e,1} - \frac{Z_0}{2}\mathbf{K}_{e,2}e^{ik_0 d} = 0, \tag{2.49}$$

for the reflection.

The general solution to the system of equations given by Eqs. (2.48) and (2.49) is

$$\mathbf{K}_{e,1} = -\frac{i}{Z_0 \sin k_0 d}\mathbf{E}_i \tag{2.50}$$

and

$$\mathbf{K}_{e,2} = \frac{i e^{-ik_0 d}}{Z_0 \sin k_0 d}\mathbf{E}_i. \tag{2.51}$$

Similar to TCMT presented in (2.2), we can fashion symmetric and asymmetric surface current densities as

$$\mathbf{K}_s = \frac{1}{2} \left(\mathbf{K}_{e,1} + \mathbf{K}_{e,2} \right) \tag{2.52}$$

and

$$\mathbf{K}_a = \frac{1}{2} \left(\mathbf{K}_{e,2} - \mathbf{K}_{e,1} \right). \tag{2.53}$$

Using symmetric and asymmetric surface current densities, our requirements for total absorption, Eqs. (2.48) and (2.49), becomes

$$-\frac{1}{2} Z_0 \mathbf{K}_s \left(1 + e^{ik_0 d} \right) - \frac{1}{2} Z_0 \mathbf{K}_a \left(1 - e^{ik_0 d} \right) = -\mathbf{E}_i, \tag{2.54}$$

for the transmission, and

$$-\frac{1}{2} Z_0 \mathbf{K}_s \left(1 + e^{ik_0 d} \right) - \frac{1}{2} Z_0 \mathbf{K}_a \left(-1 + e^{ik_0 d} \right) = 0, \tag{2.55}$$

for the reflection.

Metamaterial electromagnetic wave absorbers are typically sub-wavelength in the propagation direction ($\hat{z}$ in our case), and thus $kd \ll 1$. If we Taylor expand the exponentials in Eqs. (2.54) and (2.55) and keep only the terms necessary for parenthetical terms to be non-zero, we find for the transmission and reflection, respectively,

$$- Z_0 \mathbf{K}_s + \frac{i}{2} \mathbf{K}_a \omega \mu_0 d = -\mathbf{E}_i, \tag{2.56}$$

$$- Z_0 \mathbf{K}_s - \frac{i}{2} \mathbf{K}_a \omega \mu_0 d = 0, \tag{2.57}$$

where we have substituted $Z_0 k_0 = \omega \mu_0$. As discussed previously, the electric field may be written in terms of the wave impedance or the surface current by $E = ZH = K_m$. Thus, the first terms of Eqs. (2.56) and (2.49) are the electric fields generated by the in-phase (symmetric) total electric surface current density supported by the two sheets $\mathbf{K}_s = (\mathbf{K}_{e,1} + \mathbf{K}_{e,2})/2 = \mathbf{K}_e/2$, where $\mathbf{K}_e$ is the equivalent total electric-current sheet contributing to the electric field of the scattering from the ultrathin slab. The second terms on the left of Eqs. (2.56) and (2.57) are electric fields generated, this time, by an out-of-phase (asymmetric) total surface current density given by $\mathbf{K}_a = (\mathbf{K}_{e,2} - \mathbf{K}_{e,1})/2$, that are phase delayed ($i = \exp(i\pi/2)$) with respect to the first terms, and are proportional to the distance between sheets d. This dependence reminds us of Faraday's induction law (Eq. (2.4)), which, in integral form is, $\oint_c \mathbf{E} \cdot d\mathbf{l} = -\partial \Phi / \partial t$, where the magnetic flux is $\Phi = \int_S \mathbf{B} \cdot d\mathbf{S}$. The surface bound current is given by the asymmetric surface current as $\mathbf{K}_a = -\mathbf{M} \times \hat{n}/d$, where $\mathbf{M} = d\mathbf{m}/dV$ is the magnetization and V is the volume. Since only surface current is considered, we can show that $\mathbf{M} = \mathbf{m}/S$. With $\hat{n} = -\hat{z}$ (Fig. 2.3) and Eq. (2.47) one finds $-i\omega\mu_0 d\,\mathbf{K}_a = \hat{z} \times \mathbf{K}_m$, giving the conditions on the transmission and reflection for total absorption to be

$$-\frac{1}{2}Z_0\mathbf{K}_e + \frac{1}{2}\hat{z} \times \mathbf{K}_m = -\mathbf{E}_i, \tag{2.58}$$

$$-\frac{1}{2}Z_0\mathbf{K}_e - \frac{1}{2}\hat{z} \times \mathbf{K}_m = 0. \tag{2.59}$$

More generally, we can describe the transmission and reflection through the current sheets using dipole moments. The transmitted and reflected fields are

$$E_t = E_i - \frac{1}{2}Z_0\mathbf{K}_e + \frac{1}{2}\hat{z} \times \mathbf{K}_m, \tag{2.60}$$

$$E_r = -\frac{1}{2}Z_0\mathbf{K}_e - \frac{1}{2}\hat{z} \times \mathbf{K}_m. \tag{2.61}$$

Using Eqs. (2.46) and (2.47), we can obtain the transmission and reflection as:

$$\tilde{t} = \frac{E_t}{E_i} = 1 + \frac{iZ_0\omega}{2E_iS}\left(\mathbf{p} - \frac{\mu_0}{Z_0}\hat{z} \times \mathbf{m}\right), \tag{2.62}$$

$$\tilde{r} = \frac{E_r}{E_i} = \frac{iZ_0\omega}{2E_iS}\left(\mathbf{p} + \frac{\mu_0}{Z_0}\hat{z} \times \mathbf{m}\right). \tag{2.63}$$

Since the dipole moment $\mathbf{p} \propto E_i$ and $\mathbf{m} \propto H_i$, for convenience, we can set $E_i = 1$ V/m, so that $\mathbf{p}$ and $\mathbf{m}$ represent the dipole moments under unitary incidence. Therefore, we have:

$$\tilde{t} = 1 + \frac{ik_0}{2\epsilon_0 S}\left(\mathbf{p} - \frac{1}{c}\hat{z} \times \mathbf{m}\right), \tag{2.64}$$

$$\tilde{r} = \frac{ik_0}{2\epsilon_0 S}\left(\mathbf{p} + \frac{1}{c}\hat{z} \times \mathbf{m}\right). \tag{2.65}$$

With the assumption of zero magneto-electric coupling between the electric and magnetic dipoles, and an x-directed electric dipole (as shown in Fig. 2.3), we have

$$\tilde{t} = 1 + \frac{ik_0}{2\epsilon_0 S}\left(p_x + \frac{1}{c}m_y\right), \tag{2.66}$$

$$\tilde{r} = \frac{ik_0}{2\epsilon_0 S}\left(p_x - \frac{1}{c}m_y\right). \tag{2.67}$$

It should be noted that Eqs. (2.66) and (2.67) are the same as Eqs. (2.86) and (2.87) from multipole expansion, which we derive next.

2.4   MULTIPOLE EXPANSION

As described in Eqs. (2.64) and (2.65), the transmission and reflection coefficients are directly determined by the electric and magnetic dipole moments. However, if we have the case where higher order multipoles cannot be neglected, then the conditions for perfect absorption are modified [15]. In this section, we present more general conditions necessary to achieve perfect absorption, which include higher-order multipoles. There are two major methods used for determination of multipoles: expansion of the fields external to the particle with spherical multipoles using orthogonal spherical harmonics as basis functions [16], and using induced currents inside the particle in Cartesian coordinates based on the discrete dipole approximation [17]. Here, we adopt the later method since we would like to identify contributions to the scattering cross section due to dipoles, quadrupoles, and toroidal multipoles [18]. In the Cartesian multipole expansion method, the scattering object consists of a closely packed cubic lattice of electric dipoles with the same polarizability α_p for each dipole. All multipoles are located at the origin in Cartesian coordinates, which is also the center of mass of the disk. The continuity equation can be used to describe the induced electric dipole (ED) moment

$$\mathbf{p} = \frac{i}{\omega} \int_V \mathbf{J} d^3 \mathbf{r}, \qquad (2.68)$$

where $\mathbf{J}$ is the induced polarization current inside the resonator given by Eq. (2.45) as $\mathbf{J} = -i\omega \mathbf{P}(\mathbf{r}) = -i\omega \epsilon_0 (\epsilon_p - \epsilon_d) \mathbf{E}(\mathbf{r})$, V is the volume of the disk, $\mathbf{r}$ is the radius vector, $\epsilon_{p(d)}$ is the relative permittivity of the dielectric disk (surrounding dielectric). Here we assume that our disk resonators are embedded in air with $\epsilon_d = 1$. The magnetic dipole (MD) moment ($\mathbf{m}$), electric quadrupole (EQ) tensor ($\widehat{Q}$), magnetic quadrupole (MQ) tensor ($\widehat{M}$) and the toroidal dipole (TD) moment ($\mathbf{T}$) can be expressed as

$$\mathbf{m} = \frac{1}{2} \int_V \mathbf{r} \times \mathbf{J} d^3 \mathbf{r} \qquad (2.69)$$

$$\widehat{Q} = \frac{3i}{\omega} \int_V \left[\mathbf{r} \otimes \mathbf{J} + \mathbf{J} \otimes \mathbf{r} - \frac{2}{3} (\mathbf{r} \cdot \mathbf{J} \widehat{U}) \right] d^3 \mathbf{r} \qquad (2.70)$$

$$\widehat{M} = \frac{1}{3} \int_V \left[(\mathbf{r} \times \mathbf{J}) \otimes \mathbf{r} + \mathbf{r} \otimes (\mathbf{r} \times \mathbf{J}) \right] d^3 \mathbf{r} \qquad (2.71)$$

$$\mathbf{T} = \frac{1}{10c} \int_V \left[(\mathbf{r} \cdot \mathbf{J}) \mathbf{r} - 2r^2 \mathbf{J} \right] d^3 \mathbf{r}, \qquad (2.72)$$

where $\otimes$ denotes the dyadic product of vectors and U is the 3×3 unit tensor. Then the corresponding scattering fields due to multipoles is given as [19]

$$E(\mathbf{r})_{sca}(ED) = \frac{k_0^2}{4\pi\epsilon_0}\frac{e^{ik_0r}}{r}\left[\mathbf{n}\times(\mathbf{p}\times\mathbf{n})\right] \tag{2.73}$$

$$E(\mathbf{r})_{sca}(MD) = \frac{k_0^2}{4\pi\epsilon_0}\frac{e^{ik_0r}}{r}\left[\frac{1}{c}\mathbf{m}\times\mathbf{n}\right] \tag{2.74}$$

$$E(\mathbf{r})_{sca}(EQ) = \frac{k_0^2}{4\pi\epsilon_0}\frac{e^{ik_0r}}{r}\left[\frac{ik_0}{6}\mathbf{n}\times[\mathbf{n}\times(\widehat{Q}\mathbf{n})]\right] \tag{2.75}$$

$$E(\mathbf{r})_{sca}(MQ) = \frac{k_0^2}{4\pi\epsilon_0}\frac{e^{ik_0r}}{r}\left[\frac{ik_0}{2c}\mathbf{n}\times(\widehat{M}\mathbf{n})\right] \tag{2.76}$$

$$E(\mathbf{r})_{sca}(TD) = \frac{k_0^2}{4\pi\epsilon_0}\frac{e^{ik_0r}}{r}\left[ik_0\mathbf{n}\times(\mathbf{T}\times\mathbf{n})\right], \tag{2.77}$$

where k, c, are the wavenumber and light velocity in vacuum, $\mathbf{n} = (\sin\theta\cos\phi, \sin\theta\sin\phi, \cos\theta)^T$ is the unit vector along the radius vector $\mathbf{r}$, θ is the polar angle and ϕ is the azimuthal angle.

Without loss of generality, we assume that the resonator is independent of the polarization, and the incident field is polarized in the $\hat{x}$ direction $\mathbf{E}_{inc} = \hat{x}E_{0,x}e^{ik_0r}$, the induced polarization current $\mathbf{J}_p = (J_x, 0, 0)^T$, leads to several non-zero multipoles; p_x, m_y, $Q_{zx} = Q_{xz}$, $M_{yz} = M_{zy}$, and T_x. The scattered field in the forward direction ($\theta = 0$), and backward direction ($\theta = \pi$) can thus be described as

$$E(\mathbf{r})_{sca,x}^{+,0} \simeq \frac{k_0^2}{4\pi\epsilon_0}\frac{e^{ik_0r}}{r}\left[p_x + \frac{1}{c}m_y + \frac{k_0}{6i}Q_{zx} + \frac{k_0}{2ic}M_{zy} + ik_0T_x\right] \tag{2.78}$$

$$E(\mathbf{r})_{sca,x}^{-,\pi} \simeq \frac{k_0^2}{4\pi\epsilon_0}\frac{e^{ikr}}{r}\left[p_x - \frac{1}{c}m_y - \frac{k_0}{6i}Q_{zx} + \frac{k_0}{2ic}M_{zy} + ik_0T_x\right]. \tag{2.79}$$

According to the stationary phase approximation, we can further describe the transmitted and reflected fields in terms of the multipole polarizabilities as [20]

$$E^+ = E_{0,x}e^{ik_0z} + \iint_{-\infty}^{\infty}E_{sca,x}^+\varsigma dxdy \tag{2.80}$$

$$\simeq E_{0,x}e^{ik_0z} + \frac{i\varsigma k_0}{2\epsilon_0}e^{ik_0z}\left[p_x + \frac{1}{c}m_y + \frac{k_0}{6i}Q_{zx} + \frac{k_0}{2ic}M_{zy} + ik_0T_x\right]$$

$$= E_{0,x}e^{ik_0z}\left[1 + \frac{i\varsigma k_0}{2E_{inc}\epsilon_0}\left(p_x + \frac{1}{c}m_y + \frac{k_0}{6i}Q_{zx} + \frac{k_0}{2ic}M_{zy} + ik_0T_x\right)\right]$$

$$E^- \simeq e^{ik_0z}\frac{i\varsigma k_0}{2\epsilon_0}\left[p_x - \frac{1}{c}m_y - \frac{k_0}{6i}Q_{zx} + \frac{k_0}{2ic}M_{zy} + ik_0T_x\right], \tag{2.81}$$

where $z \to \infty$ is the propagation distance from the center of the cylinder in the z direction and $\varsigma = 1/S$ is the density of the multipoles per unit area.

If we take the reference plane at z, then the transmission and reflection coefficient can be written as

$$\tilde{t} = \frac{E^+}{E_{0,x}} = 1 + \frac{ik_0\varsigma}{2E_{0,x}\epsilon_0}\left(p_x + \frac{1}{c}m_y + \frac{k_0}{6i}Q_{zx} + \frac{k_0}{2ic}M_{zy} + ik_0T_x\right) \qquad (2.82)$$

$$\tilde{r} = \frac{E^-}{E_{0,x}} = \frac{ik_0\varsigma}{2E_{0,x}\epsilon_0}\left(p_x - \frac{1}{c}m_y - \frac{k_0}{6i}Q_{zx} + \frac{k_0}{2ic}M_{zy} + ik_0T_x\right). \qquad (2.83)$$

The condition of perfect absorption dictates that the transmission and reflection must be zero simultaneously. We can obtain the relations with the assumption of $E_{0,x} = 1$ V/m:

$$\text{Re}\left[p_x + \frac{k_0}{2ic}M_{zy} + ik_0T_x\right] = \text{Re}\left[\frac{1}{c}m_y + \frac{k_0}{6i}Q_{zx}\right] = 0 \qquad (2.84)$$

$$\text{Im}\left[p_x + \frac{k_0}{2ic}M_{zy} + ik_0T_x\right] = \text{Im}\left[\frac{1}{c}m_y + \frac{k_0}{6i}Q_{zx}\right] = \frac{\epsilon_0}{k_0\varsigma}. \qquad (2.85)$$

If only electric dipole and magnetic dipole components are considered to contribute the transmission and reflection, we can get the same conditions as shown in Eqs. (2.64)–(2.65):

$$\tilde{t} = 1 + \frac{ik_0}{2\epsilon_0 S}\left(p_x + \frac{1}{c}m_y\right) \qquad (2.86)$$

$$\tilde{r} = \frac{ik_0}{2\epsilon_0 S}\left(p_x - \frac{1}{c}m_y\right). \qquad (2.87)$$

In order to achieve $\tilde{t} = \tilde{r} = 0$, i.e., perfect absorption, the electric and magnetic dipole moments must satisfy

$$\text{Re}\{p_x\} = \frac{1}{c}\text{Re}\{m_y\} = 0 \quad \text{and} \quad \text{Im}\{p_x\} = \frac{1}{c}\text{Im}\{m_y\} = \frac{\epsilon_0 S}{k_0}. \qquad (2.88)$$

2.5 INTERFERENCE THEORY OF PERFECT ABSORBERS

In 2012, a theory of metamaterial absorbers based on interference was published and described an alternative view not based on effective medium theory [21]. In this approach, the two metal layers are considered to be decoupled such that multiple wave reflections occurring inside the finite-thickness spacer layer. The method is thus similar to that of the Salisbury and Jaumann absorbers described in Section 1.2, where the interference condition for absorption is given as $d = \lambda_0/4n$.

As shown in Fig. 2.3, the top metamaterial layer is assumed to be an infinitesimal impedance surface. As the electromagnetic waves is incident onto the air-metal interface, reflection and transmission take place with a reflection coefficient $\tilde{r}_{12} = r_{12}e^{i\phi_{12}}$ and transmission coefficient $\tilde{t}_{12} = t_{12}e^{i\theta_{12}}$. The transmitted wave continues propagating inside the spacer layer with a propagation wavenumber $\tilde{k} = \sqrt{\tilde{\epsilon}_d}k_0$, where $\tilde{\epsilon}_d$ is the permittivity of the dielectric spacer layer of thickness d. The wave reflects back from the conducting ground plane with a reflection coefficient $\tilde{r}_{23}$. Multiple reflection and transmission may occur at the metal-air interface, with coefficients $\tilde{r}_{21} = r_{21}e^{i\phi_{21}}$ and $\tilde{t}_{21} = t_{21}e^{i\theta_{21}}$. Finally, the total reflection out of the slab is the superposition of the direct reflection from the metamaterial layer and the multiple reflections at the air-metal interface:

$$\tilde{r} = \tilde{r}_{12} + \frac{\tilde{t}_{12}\tilde{t}_{21}}{1 - \tilde{r}_{21}\tilde{r}_{23}e^{i2\tilde{k}d}}\tilde{r}_{23}e^{i2\tilde{k}d}. \tag{2.89}$$

For a ground-plane supported perfect absorber the transmission is zero, and therefore the absorbance is given by $A = 1 - |\tilde{r}|^2$. Since the top metamaterial layer is decoupled from the bottom ground plane, the transmission and reflection coefficients can be easily obtained by removing the ground plane layer and setting the metamaterial on a half dielectric space. For a given metamaterial structure and spacer material, the overall reflection is only determined by the thickness of spacer d, and according to [21] the total round-trip phase is $2kd + \phi_{21} + \pi \approx 2\pi$. Therefore if $\phi_{21} = 0$—or is a multiple of 2π—we find $d = \lambda_0/4n$.

2.6 DEPENDENCE OF ABSORPTION BANDWIDTH ON THICKNESS

For a ground plane-supported absorber, the bandwidth of absorption is determined by the sum of the radiative and material loss rates, i.e.,

$$BW \equiv 2\left|\frac{\omega - \omega_0}{\omega}\right| = \frac{2(\gamma + \delta)}{\omega_0} \tag{2.90}$$

according to Eq. (2.41). The radiative loss rate is dominated by the geometry of the resonator, while the material loss rate is not only regulated by the loss tangents of the constituents, but is also related to the total thickness of the absorber.

Based on the Fresnel equations, previous analytical study on a single-layer absorber has shown that the bandwidth is bounded as [22]

$$BW \approx \frac{4c}{\omega d}\frac{\rho}{|\mu_r - \epsilon_r|}, \tag{2.91}$$

where ρ is the reflection coefficient at ω, which is close to the perfect absorption frequency, and μ_r, ϵ_r are the relative permittivity and permeability of the slab. Equation (2.91) shows that the bandwidth of a single-layer absorber is inversely proportional to the absorber thickness, which is contradictory to the analysis in [23], which shows that the bandwidth increases with the increasing thickness. This inconsistency arises from the fact that the d in Eq. (2.91) is the thickness for a perfect absorber, which is also determined by the permittivity and permeability. In addition, ϵ and μ in the analysis were assumed to be dispersiveless, which is not true since the real and imaginary parts of them are correlated by the Kramers–Kronig relations. Otherwise, the absorber will be broadband with an infinite bandwidth.

A more fundamental relationship between the bandwidth and the absorber thickness can be derived for any single-layer metal backed absorber. Another assumption is that the reflection coefficient can be described keeping only the first-order term in λ^{-1}, which is given as

$$\tilde{r}(\lambda)|_{\lambda \to \infty} = -1 - i \frac{4\pi \mu d}{\lambda}. \tag{2.92}$$

Using the reflectivity coefficient given in Eq. (2.92), a Kramers–Kronig analysis shows that the absorber thickness is given by [24],

$$d \geq \frac{1}{2\pi^2 \mu_s} \left| \int_0^\infty \ln |\tilde{r}(\lambda)| \, d\lambda \right|, \tag{2.93}$$

where $\mu_s = Re\{\mu\}|_{\lambda \to \infty}$ is the static permeability of the absorber.

2.7   SUMMARY

In this chapter, we have presented several theoretical approaches which describes the electromagnetic properties of perfect absorbers. In addition to the theories detailed here, there are two other common approaches—transmission-line theory, and the equivalent-circuit theory, both of which are prevalent in the field of microwave engineering. There are several good books and papers covering these additional methods [11, 25–29].

2.8   REFERENCES

[1] John David Jackson. *Classical Electrodynamics*, 3rd ed., Wiley, New York, NY, 1999. 8

[2] Peter Markoš and Costas M. Soukoulis. *Wave Propagation: From Electrons to Photonic Crystals and Left-Handed Materials*. Princeton University Press,—student edition, 2008. DOI: 10.1515/9781400835676 10

[3] S. Fan, W. Suh, and J. D. Joannopoulos. Temporal coupled-mode theory for the fano resonance in optical resonators. *Journal of the Optical Society of America A*, 20(3):569–572, March 2003. DOI: 10.1364/josaa.20.000569 13

[4] W. Suh, Z. Wang, and S. Fan. Temporal coupled-mode theory and the presence of non-orthogonal modes in lossless multimode cavities. *IEEE Journal of Quantum Electronics*, 40(10):1511–1518, October 2004. DOI: 10.1109/jqe.2004.834773 13

[5] M. F. Yanik, S. Fan, M. Soljačić, and J. D. Joannopoulos. All-optical transistor action with bistable switching in a photonic crystal cross-waveguide geometry. *Optics Letters*, 28(24):2506–2508, December 2003. DOI: 10.1364/ol.28.002506 13

[6] Y. Ra'di, C. R. Simovski, and S. A. Tretyakov. Thin perfect absorbers for electromagnetic waves: Theory, design, and realizations. *Physical Review Applied*, 3(3):037001, March 2015. DOI: 10.1103/physrevapplied.3.037001 14, 17

[7] Chloe F. Doiron and Gururaj V. Naik. Non-hermitian selective thermal emitters using metal–semiconductor hybrid resonators. *Advanced Materials*, 31(44):1904154, September 2019. DOI: 10.1002/adma.201904154 14

[8] Shih-Hui Gilbert Chang and Chia-Yi Sun. Avoided resonance crossing and non-reciprocal nearly perfect absorption in plasmonic nanodisks with near-field and far-field couplings. *Optics Express*, 24(15):16822, July 2016. DOI: 10.1364/oe.24.016822 14

[9] H. A. Wheeler. The radiation resistance of an antenna in an infinite array or waveguide. *Proc. of the IRE*, 36(4):478–487, April 1948. DOI: 10.1109/jrproc.1948.229650 16

[10] H. Wheeler. Simple relations derived from a phased array made of an infinite current sheet. In *1964 Antennas and Propagation Society International Symposium*. Institute of Electrical and Electronics Engineers, 1964. DOI: 10.1109/aps.1964.1150148 16

[11] Ben A. Munk. *Finite Antenna Arrays and FSS*. John Wiley & Sons, Inc., July 2003. DOI: 10.1002/0471457531 16, 25

[12] R. C. Hansen. Linear connected arrays. *IEEE Antennas and Wireless Propagation Letters*, 3:154–156, 2004. DOI: 10.1109/lawp.2004.832125 17

[13] S. A. Tretyakov, A. J. Viitanen, S. I. Maslovski, and I. E. Saarela. Impedance boundary conditions for regular dense arrays of dipole scatterers. *IEEE Transactions on Antennas and Propagation*, 51(8):2073–2078, August 2003. DOI: 10.1109/tap.2003.814737 17

[14] Sergei Tretyakov. *Analytical Modeling in Applied Electromagnetics*. Artech House, Boston, 2003. 17

[15] J. Y. Suen, K. Fan, and W. J. Padilla. A zero-rank, maximum nullity perfect electromagnetic wave absorber. *Advanced Optical Materials*, page 1801632, January 2019. DOI: 10.1002/adom.201801632 21

[16] P. Grahn, A. Shevchenko, and M. Kaivola. Electromagnetic multipole theory for optical nanomaterials. *New Journal of Physics*, 14(9):093033, September 2012. DOI: 10.1088/1367-2630/14/9/093033 21

[17] Pavel D. Terekhov, Kseniia V. Baryshnikova, Yuriy A. Artemyev, Alina Karabchevsky, Alexander S. Shalin, and Andrey B. Evlyukhin. Multipolar response of nonspherical silicon nanoparticles in the visible and near-infrared spectral ranges. *Physical Review B*, 96:035443, July 2017. DOI: 10.1103/physrevb.96.035443 21

[18] Andrey E. Miroshnichenko, Andrey B. Evlyukhin, Ye Feng Yu, Reuben M. Bakker, Arkadi Chipouline, Arseniy I. Kuznetsov, Boris Luk'yanchuk, Boris N. Chichkov, and Yuri S. Kivshar. Nonradiating anapole modes in dielectric nanoparticles. *Nature Communications*, 6(1):8069, 2015. DOI: 10.1038/ncomms9069 21

[19] Jun Chen, Jack Ng, Zhifang Lin, and C. T. Chan. Optical pulling force. *Nature Photonics*, 5(9):531–534, 2011. DOI: 10.1038/nphoton.2011.153 22

[20] Anders Pors, Sebastian K. H. Andersen, and Sergey I. Bozhevolnyi. Unidirectional scattering by nanoparticles near substrates: generalized Kerker conditions. *Optics Express*, 23(22):28808, November 2015. DOI: 10.1364/oe.23.028808 22

[21] Hou-Tong Chen. Interference theory of metamaterial perfect absorbers. *Optics Express*, 20(7):7165, March 2012. DOI: 10.1364/oe.20.007165 23, 24

[22] George T. Ruck, Donald E. Barrick, and Stuart William. *Radar Cross Section Handbook*, 1st ed., Peninsula Publishing, 1970. DOI: 10.1007/978-1-4899-5324-7 24

[23] Claire M. Watts, Xianliang Liu, and Willie J. Padilla. Metamaterial electromagnetic wave absorbers. *Advanced Materials*, 24(23):OP98–OP120, May 2012. DOI: 10.1002/adma.201200674 24

[24] K. N. Rozanov. Ultimate thickness to bandwidth ratio of radar absorbers. *IEEE Transactions on Antennas and Propagation*, 48(8):1230–1234, 2000. DOI: 10.1109/8.884491 25

[25] Christophe Caloz and Tatsuo Itoh. *Electromagnetic Metamaterials: Transmission Line Theory and Microwave Applications*. John Wiley & Sons, Inc., November 2005. DOI: 10.1002/0471754323 25

[26] Ricardo Marqus, Ferran Martn, and Mario Sorolla. *Metamaterials with Negative Parameters*. John Wiley & Sons, Inc., December 2007. DOI: 10.1002/9780470191736 25

[27] Filippo Costa, Simone Genovesi, Agostino Monorchio, and Giuliano Manara. A circuit-based model for the interpretation of perfect metamaterial absorbers. *IEEE Transactions on Antennas and Propagation*, 61(3):1201–1209, March 2013. DOI: 10.1109/tap.2012.2227923 25

[28] Filippo Capolino. *Theory and Phenomena of Metamaterials*. CRC Press, December 2017. DOI: 10.1201/9781420054262 25

[29] G. Duan, J. Schalch, X. Zhao, J. Zhang, R. D. Averitt, and X. Zhang. Identifying the perfect absorption of metamaterial absorbers. *Physical Review B*, 97(3):035128, January 2018. DOI: 10.1103/physrevb.97.035128 25

CHAPTER 3

Metamaterial Perfect Absorbers and Performance

Since the first experimental demonstration of the electromagnetic perfect absorber by Landy et al. in 2008 [1], studies of metamaterial perfect absorbers (MPAs) have traversed the electromagnetic spectrum from radio frequencies to the visible range. Over the past years, various novel MPA designs and functionalities—which can be categorized into plasmonic or all-dielectric based MPAs—have been proposed and demonstrated. The MPA design is extremely versatile thus permitting great flexibility in performance and applications. For example, metamaterial perfect absorbers have been shown useful for chemical sensing, solar energy harvesting, imaging, detection, thermophotovoltaics, and color filtering. In this chapter, we will review the development of metamaterial absorbers over the past decade. After describing the general concept of the metamaterial perfect absorber, we will discuss the main features of MPAs, including polarization performance, broad angular acceptance, broad bandwidth, and spatial dependence.

3.1 INTRODUCTION

Electromagnetic metamaterials often consist of arrays of patterned subwavelength elements whose electromagnetic response can be described using effective electric permittivity ε and magnetic permeability μ. Through engineering the metamaterial geometry, impedance matching to free space can be achieved with the condition of $\varepsilon_r(\omega) = \mu_r(\omega)$, which permits zero reflection from the metamaterial. MPAs have been shown to outperform other well-known resonant absorbers, such as the Salisbury screen and Dällenbach absorber [2, 3] in terms of the scalability, form factor, angular dependence, bandwidth and tunability. For metallic metamaterial absorbers, the inclusion of a ground plane ensures no transmission through the structure, thus leading to near unity absorption. The absorbed power is primarily dissipated inside the lossy dielectric spacer, lying in-between the metamaterial layer and ground plane [1]. However, for all-dielectric metamaterials—under the condition of impedance matching—perfect absorption is realized through achieving degenerate hybrid electric and hybrid magnetic modes, and selecting a material dielectric loss tangent, such that the power is directly absorbed inside the dielectric metamaterials [4, 5]. Since both the metamaterial element and periodicity size are subwavelength, incident angles for MPAs are much larger than those for FSSs [6].

Since the first experimental demonstration of the metamaterial perfect absorber (MPA) in 2008, there has been intense interest in development of practical applications. With the development of MPAs over the last decade, researchers have experimentally shown designs operating over nearly the entire electromagnetic spectrum from radio frequency, millimeter wave, terahertz, infrared, through optical to ultraviolet range. Figure 3.1 shows some milestones of the past progress of MPAs, which covers various major topics, over the years from 2002–2021.

In what follows, we categorize MPAs into three major types: metallic MPAs (MMPAs), all-dielectric MPAs (DMAs), and coherent perfect absorbers (CPAs). In Section 3.2, we will first review the conventional configuration of MPAs based on metals. Then, we will discuss the state-of-the-art of DMAs and CPAs, respectively. In Section 3.3, we will discuss the primary electromagnetic properties that can be modified with MPAs, including bandwidth, polarization, independence, and angular acceptance. In Section 3.4, we will discuss the general geometries and materials used for MPA designs.

3.2   METAMATERIAL PERFECT ABSORBERS

3.2.1   METALLIC METAMATERIAL PERFECT ABSORBERS

The first metamaterial based perfect absorber by Landy et al. as shown in Fig. 3.2a, demonstrated a simulated absorptivity of 99% at 11.48 GHz [1]. The top layer is an ERR which provides the electric response by coupling strongly to the incident electric field at a desired resonance frequency. The second metal, spaced apart from the top layer by a dielectric, consists of a cut wire in a parallel plane, and also contributes to the electric response. Magnetic coupling is achieved via circulating displacement currents in the cut wire and the center wire of the ERR. An incident time-varying magnetic field may couple to these anti-parallel currents, thus yielding a Lorentz-like magnetic response. The combined design allows for individual tuning of the electric and magnetic responses. For example, adjustment of the geometry of the ERR permits tuning the frequency position and strength of the resonance, while altering the spacing of the two metallic structures, and their geometry, allows the magnetic response to be modified. As a result, the effective impedance at resonance, out of which effective permittivity and effective permeability are induced by the two orthogonal circulating current and circulating displacement current, matches the free space impedance [12]. As shown in Fig. 3.2c–d, the authors showed that the experimentally measured transmissivity and reflectivity are very close to the simulated results.

The initial MMPA requires design of both of the top and bottom resonators to achieve impedance matching [1, 54]. To reduce the parameter design space and fabrication complexity, the bottom metal layer was replaced directly by a metallic ground plane, working as a PEC layer, i.e., a mirror plane for the electric field, to ensure negligible transmission through the absorber [8]. Based the TCMT analysis presented in Chapter 2, the ground plane MMPA may be described as a single-port system. Through optimization of the distance between the resonator and the ground plane, the reflection from the front of the resonator can be cancelled via destructive interference of the reflection and incoming waves. It is clear that as long as the resonator is critically coupled with intrinsic loss rate balancing the radiative decay rate, the reflection coefficient becomes zero, indicating a perfect absorption state.

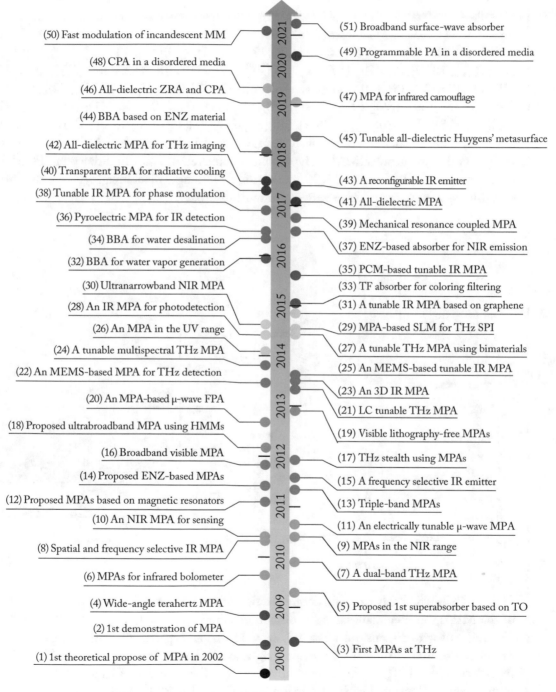

Figure 3.1: (*Continues.*)

Figure 3.1: (*Continued.*) Milestones of metamaterial perfect absorber since 2002. (1)[7] (2)[1], (3)[8], (4)[6], (5)[9], (6)[10], (7)[11],(8)[12], (9)[13], (10)[14],(11)[15], (12)[16], (13)[17], (14)[18], (15)[19], (16)[20], (17)[21], (18)[22], (19)[23], (20)[24], (21)[25], (22)[26], (23)[27], (24)[28], (25)[29], (26)[30], (27)[31], (28)[32], (29)[33], (30)[34], (31)[35], (32)[36], (33)[37], (34)[38], (35)[39], (36)[40], (37)[41], (38)[42], (39)[43], (40)[44], (41)[4],(42)[5], (43)[45], (44)[46], (45)[47], (46)[48], (47)[49], (48)[50], (49)[51], (50)[52], (51)[53]. BBA - broadband absorber, CPA - coherent perfect absorber, ET - electrically tunable, ENZ - epsilon near zero, FPA - focal plane array, HMM - Hyperbolic metamaterial, LC - liquid crystal, PCM - phase change material, SLM - spatial light modulator, SPI - single pixel imaging, TF - thin film, TO - transformation optics, ZRA - zero-order rank absorber

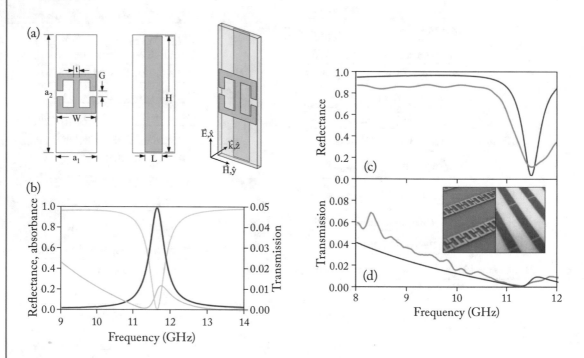

Figure 3.2: The first experimental demonstration of a metamaterial perfect absorber. (a) Unit cell of the MPA: top metamaterial layer (dimensions in mm: $a_1 = 4.2$, $a_2 = 12$, $W = 4$, $G = 0.6$, $t = 0.6$), cut wire dimensions ($L = 1.7$, $H = 11.8$), and a perspective view of the unit cell. (b) Simulated results of transmissivity (blue, right axis), reflectivity (green, left axis), and absorptivity (red, left axis). (c,d) Experimentally measured (blue) and simulated (red) reflectivity and transmissivity. Reprinted with permission from Ref. [1] © 2008 APS.

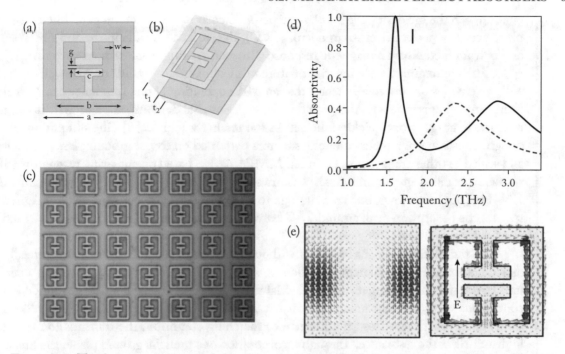

Figure 3.3: Terahertz metamaterial absorber consisting of two metallic layers and two dielectric layers. (a) Electric SRR: unit cell a: 36 μm, SRR side length b: 25.9 μm, capacitor length c: 10.8 μm, capacitor gap g: 1.4 μm, and linewidth w:3 μm. (b) Perspective view of the absorber. Each dielectric layer, t1 and t2, is 8 μm thick. (c) Photograph of a portion of the experimentally realized absorber. (d) The simulated absorptivity (for normal incidence) out to 3.5 THz revealing an additional absorption. The red dashed curve shows the absorption with the capacitor shorted highlighting the importance of the in-plane electric response in determining the main absorption peak. (e) The calculated surface current density at the absorption peak. Reprinted with permission from Ref. [6] © 2008 APS.

The first GPA was experimentally demonstrated at the terahertz frequencies by Tao et al. in 2008 [6]. The structure consists of a split-ring resonator (SRR) on top of an 8-μm polyimide spacer and a 200-nm gold continuous ground plane. Through geometrical optimization, the absorptivity shows a peak at 1.6 THz; see Fig. 3.2d. Similar to the working principle of the first MPA, the perfect absorptivity is achieved through exciting the electrical response in the top SRR structure and the magnetic response between two metal layers. Therefore, the effective impedance at the resonance matches the free space and the total incident power is absorbed inside the metamaterial layer. Figure 3.2e shows the calculated surface currents in the top resonant structure and the lower ground plane at the peak absorption frequency. The surface currents are associated with an LC resonance, and can be seen to counter-circulate in the top electric resonator when the incident wave is polarized perpendicular to the resonator gap. Since the ground plane acts as a perfect electric conductor (PEC), the surface currents in the top ring resonator and bottom metal form a magnetic response with a circulating displacement current.

The approximate scale invariant nature of metamaterials—valid for Maxwell's equations with no charges or currents—ensures that an MPA operating at one frequency can be scaled to nearly any other frequency. Although the first experimental demonstration on MPAs was in the microwave range, work quickly followed in the terahertz regime [6]. The ubiquitous MMPA design, consisting of a metamaterial resonators patterned on top of ground plane, was adopted and scaled to other frequencies. A similar MMPA design based on split-ring resonators (SRRs) was soon realized in the millimeter wave range with resonant frequency of approximately at 77 GHz [55]. As shown in Fig. 3.4a, due to the polarization dependent electromagnetic response, the absorption peak around 77 GHz only occurs when the electric field is aligned perpendicular to the resonator gap.

In the infrared range, MMPAs without the ground plane had been theoretically predicted before experimental demonstration. A wide-angle infrared absorber, based on a perfectly impedance matched negative-index material with double-layer metal strips surrounded by two continuous wires, exhibits an absorptivity of 90% at 1.5 μm [57]. Although zero reflection is achieved in this design from the impedance matching condition, the transmission remains as high as 10% at the resonance. Through incorporation of a metallic ground plane, the first experimental demonstration of MMPAs exhibited 97% absorptivity at wavelength of 6 μm [12]. As shown in Fig. 3.4b, different from the ERR structure in the microwave and terahertz ranges, the unit cell consists of arrayed cross-shaped metamaterials on a 185-nm thin dielectric layer of Al_2O_3. The retrieved optical constants $\varepsilon(\omega)$ and $\mu(\omega)$ both cross zero at the resonant frequency which ensures zero reflection with impedance matching. At other frequencies, either $\varepsilon_1(\omega)$ or $\mu_1(\omega)$ is negative such that no energy can be transmitted through the absorber. Further simulation also confirmed that most of the energy is dissipated in the dielectric layer rather than by the Ohmic loss, which is in agreement with other prior work [54]. The absorbed power subsequently is converted to local heat, that can be used for bolometric detection [10, 54, 58–60], pyroelectric detection [40, 61–63], photothermal reshaping [64], water desalination [36, 38], and energy harvesting [65].

The first two MMPAs working at NIR wavelengths [13, 14], published within days of each other, utilized a gold disk and rectangular patch in a unit cell patterned on a dielectric layer, respectively. As shown in Fig. 3.4c, Hao et al. showed that the patterned MMPAs realized absorptivity of 88% at wavelength of 1.6 μm. The absorption was explained by the excitation of plasmonic electric resonance and localized magnetic resonance between the top and bottom metals. Different from MMPAs operating at longer wavelength, the major absorbed power was dissipated as Ohmic loss.

At even shorter wavelengths, i.e., the visible and UV ranges, it is challenging to produce MMPAs, owing to limits of conventional lithography techniques. In addition, the operational

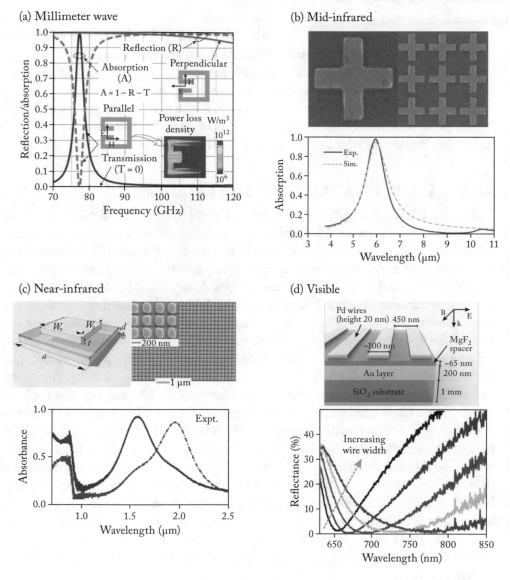

Figure 3.4: Experimental demonstration of MMPAs at various frequency ranges. (a) Experimental demonstration of MMPA in the millimeter wave range and unit cell. (b) First experimental demonstration of MMPA in the mid-infrared range and the SEM image fabricated samples. (c) One of the first experimental demonstration of MMPA in the NIR range and the SEM image of fabricated samples. (c) First experimental demonstration of MMPAs in the visible range for Hydrogen sensing and the configuration. Reprinted with permission from: (a) Ref. [55] © 2011 AIP Publishing; (b) Ref. [12] © 2010 APS; (c) Ref. [13] © 2010 AIP Publishing; (d) Ref. [56] © 2011 ACS.

frequencies are very close to plasmonic frequencies of noble metals such as gold, silver, and aluminum, indicating a large imaginary permittivity based on the Drude model [66, 67]. Therefore, it is challenging to obtain narrowband MMPAs in this realm. There are only a few experimental work with narrowband absorption demonstrated in this range, in contrast to those in longer wavelengths [56]. Although it is difficult to obtain narrow-band absorptivity at shorter wavelengths, broadband absorbers are indeed possible. The majority of studies of MMPAs in this regime possess relatively large bandwidth, and operate under different mechanisms, such as interference, and will be addressed in Section 3.3.1.

3.2.2   ALL-DIELECTRIC METAMATERIAL PERFECT ABSORBERS

In Section 3.2, we have shown that traditional metal-based MMs obtain their electromagnetic properties from individual sub-wavelength unit cells, and relatively complex response may be achieved through a bottom up approach [1, 68, 69]. Metal-based MPAs have shown polarization dependent and independent response, broadband behavior, angular tuning, precise wavelength dependence [1, 70, 71], increased effective absorption coefficient [4], and tailored spatial-dependent response [19, 72]. Traditional MPAs, however, are based on metals and as such, one often chooses the highest electrically conducting materials in order to maximize performance. Indeed the frequency dependence of the conductivity—and its subsequent reduction in the infrared and visible range—is what limits MMs from shorter wavelength operation. Thus, the performance of traditional MPAs is strongly tied to the electrical conductivity. For example, at NIR and visible wavelengths, less conductive refractory metals like tungsten (W), nickel (Ni), and molybdenum (Mo) can be used for broadband MPAs but with limited wavelength selectivity. Although this may be a suitable dependence for many areas of investigation [29], it nonetheless presents a restriction that may limit potential applications. The ratio of the electronic contribution of the thermal conductivity (κ) to the electrical conductivity (σ) is a constant and independent of any particular metal. This fundamental relationship—termed as the Wiedemann–Franz law—is usually characterized by the Lorenz number given as the ratio $L = (\kappa/\sigma T)$, which is found to be independent of any material constants, but rather is a ratio of universal constants $L = (\pi k_B/e)^2/3$. Thus, for MPA studies where thermal properties are important, the availability of a metal-free substitute would enable an alternative and less fundamentally restrictive design approach.

In the visible range, maximizing the efficiency of solar cells requires a relatively thick silicon layer to maximize the absorptivity. However, to maximize current collection, one desire to possess thinner silicon to reduce unnecessary carrier loss from scattering and electron-hole recombination. Although metal-dielectric-metal-based MPAs have been used to solve such a contradiction, the introduced metal dissipates a significant portion of the absorbed power, which is converted into heat, rather than for the generation of photocarriers. Therefore, new ultrathin metal-free structures with near perfect absorption would enable new breakthrough for energy harvesting.

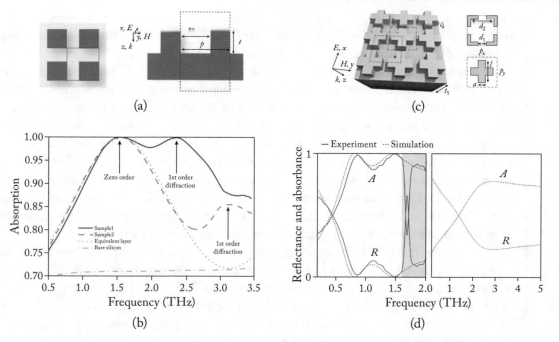

Figure 3.5: (a) The front and side views of the structure. The rectangular region is the unit cell used in simulations. (b) Absorption spectra of samples with different periods. The cases for a bare doped silicon slab (dash-dot cyan) and an absorber (dotted black) based on quarter-wavelength antireflection layer are also shown. (c) Schematic of a 2D array carved from a doped silicon substrate. The geometric parameters are: $p_x = p_y = 200\,\mu m$, $t_b = 65\,\mu m$, $a = 60\,\mu m$, $l = 160\,\mu m$, and $t_s = 200\,\mu m$. (d) Numerically (dotted) and experimentally (solid) resolved spectra. Left: Reflectance $R(\omega)$ and absorbance $A(\omega)$ for the 2D cross array absorber. Right: Reflectance $R(\omega)$ and absorbance $A(\omega)$ for the bare, doped silicon substrate. The shaded area marks the limit of the system dynamic range. Reprinted with permission from Refs. (a)–(b) [73] © 2012 OSA, (c)–(d) [74] © 2015 Wiley-VCH.

All-dielectric material based absorbers were initially proposed and demonstrated as a broadband absorber with subwavelength structures patterned on a thick and absorptive substrate [73–78]. In 2012, Luo and co-workers proposed a broadband terahertz absorber based on patterned highly doped silicon posts on a thick silicon substrate with resistivity of $0.77\,\Omega \cdot cm$ as shown in Fig. 3.5a [73]. The numerically simulated absorptivity exhibits two absorption peaks at 1.5 THz and 2.3 THz as shown in Fig. 3.5b, mainly arising from zero—and first-order diffraction, respectively. Broadband absorption was obtained by using both the zero-order and the first-order diffraction modes. The zero-order diffraction corresponds to the impedance matching condition with a quarter-wave thickness of an equivalent medium, while the first-order results from grating diffraction inside the silicon substrate. When the carrier density is much higher, with a wafer resistivity between 0.02 and 0.05 $\Omega \cdot$ cm, the cross structures shown in Fig. 3.5c, patterned directly out of silicon, behave more like plasmonic resonators which are similar to those made of metals but with much higher material loss [74]. Similarly, the measured and numerically simulated absorptivity spectra also exhibit two absorption peaks around 0.86 THz and 1.49 THz, as shown in Fig. 3.5d. Because of highly conductive silicon, however, the incident terahertz wave can excite localized surface plasmon resonances (LSPRs), in which the lower-frequency resonance is attributed to the small cavity between the top and bottom arms, while the high-frequency resonance originates from the large square cavity formed by the four adjacent resonators. For these broadband absorbers, a significant portion of the absorbed power is dissipated inside the lossy substrate, which shows an absorptivity as large as 70% without any patterning [73–75]. Therefore, a challenge with such designs is obtaining frequency selective narrowband absorption using patterned structures directly on a lossy substrate.

Recently, it has been reported that a metasurface can be fashioned entirely from sub-wavelength dielectric resonators [74, 79–81], which act as isolated "metamaterial atoms." High-index dielectric cylinders with finite height and radius have been found to support electric and magnetic resonances, which can be controlled by modifying the aspect ratio [82]. Through overlapping the two resonances, the lossless all-dielectric metasurface behaves as a Huygens' metasurface with effective impedance matching to the free space without reflection [47, 83, 84]. When introduced with adequate loss, no light is transmitted, but rather is completely absorbed entirely within the dielectric. The resulting composite material thus functions as a sub-wavelength metasurface with tunable permittivity and permeability permitting the realization of an all-dielectric absorber [85].

It was shown that the dielectric cylinders can be modeled as dielectric waveguides, which support hybrid HE and EH modes. If the z axis is oriented parallel to the cylindrical axis, then $H_z/E_z \ll 1$ for the HE mode, and $E_z/H_z \ll 1$ for the EH mode [86]. Based on the cylindrical geometry, the effective hybrid magnetic (HE_{111}) and electric (EH_{111}) dipole modes can be tuned independently. As shown in Fig. 3.6a–b, for an array of dielectric cylinders, both the HE_{111} and EH_{111} modes decrease with the radius and height of the cylinder increase. Through tuning the geometry, impedance matching can be achieved at the resonant frequency where the HE_{111} and EH_{111} modes overlap. For materials with relatively high dielectric constants ($\varepsilon_r \geq 10$), the surfaces can be approximated as perfect magnetic walls. Under these conditions, the dimensions of the dielectric resonator are determined only by the free-space resonance wavelength (λ_0) and refractive index (n) of cylinders [4]. The HE magnetic dipole mode has a cutoff given by $\lambda = h/2$, where $\lambda = \lambda_0/n$ and h is the cylinder height. Thus, given $\lambda = h/2$, the critical criterion is to find the cutoff conditions for the EH mode. Prior studies have shown that for a cylindrical dielectric particle with an index of refraction of n, radius r, and height h, the EH cutoff condition is given by

$$r = 0.61 \frac{\lambda_0}{(n^2 - 1)^{1/2}}. \tag{3.1}$$

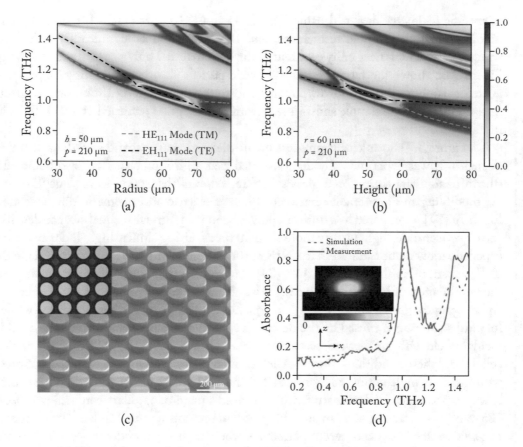

Figure 3.6: (a) Dependence of absorbance for HE_{111} (gray dashed curves) and EH_{111} (black dashed curves) cylindrical modes as a function of radius with h = 50 μm and p = 210 μm. (b) Dependence of the absorbance on frequency and height with r = 60 μm and p = 210 μm. (c) Oblique view of the fabricated arrayed silicon cylinders on a PDMS substrate. The scale bar is 200 μm. The simulated (dashed curve) and measured (solid curve) absorbance for the absorber with dimension of cylinder height = 50 μm, radius = 62.5 μm, and periodicity of 172 μm. The inset shows the on-resonance power loss density at the mid plane. Reprinted with permission from Ref. [4] © 2017 OSA.

Then given the condition for the HE mode,

$$h = \frac{\lambda_0}{2n}. \tag{3.2}$$

A specific geometry designed using Eqs. (3.1) and (3.2), guarantees the overlap of magnetic dipole and electric dipole resonances, and the absorption is significantly enhanced in the structure—reaching near unity—although the bulk material only possesses relatively small loss. Figure 3.6c shows the fabricated all-dielectric perfect absorber consisting of periodic silicon cylinders supported by a PDMS polymer substrate. The terahertz time-domain spectroscopic measurement on the sample shows a nearly unity absorption achieved at 1.011 THz, as shown in Fig. 3.6d.

Figure 3.7 shows recent progress on all-dielectric narrowband absorbers at different frequencies. It was experimentally demonstrated that a NIR metasurface absorber, consisting of silicon nanopillars sitting on a glass substrate, exhibits absorption as high as 95% at 855-nm via modifying the interference between coherent electric and magnetic dipole scattering; see Fig. 3.7a [87]. Compared to a uniform poly-silicon thin film with equal surface density, the absorber concentrates light energy more than 50 times. Using similar high-index materials, it was experimentally demonstrated that a NIR absorber can be obtained using arrayed Ge disks on a CaF$_2$ substrate, as shown in Fig. 3.7b [88]. At the absorption peak of approximately 1200-nm, the annihilation of backward scattering arises from the establishment of equal effective electric and magnetic polarizabilities, i.e., $\alpha_{eff}^e = \alpha_{eff}^m$, known as the first Kerker Condition while the zero forward scattering is caused by the destructive interference between the scattered field and the incident field with the forward-scattering cross section equal to the physical cross section of unit cell. Under these conditions, the all-dielectric metasurface can sustain high absorption over 80% with incident angle within 28° for both TE and TM polarizations. Based on the same physics, it was experimentally demonstrated that arrayed square nanopillars can achieve absorption at 785-nm [89]. Actually, such an absorbing system presents a more fundamental absorber, which can be described as two coherent perfect absorbers with eigenmode of opposite symmetry, i.e., even (EH) mode and odd (HE) mode. Based on temporal coupled-mode theory analysis, it was shown that the rank of the scattering matrix for each coherent absorption mode is unity, and that for the perfect absorber with two coherent absorption modes is zero [48]. Therefore, the all-dielectric absorber can also be described as a zero-rank absorber. The absorption of the zero-rank absorber at resonance is independent on the phase difference between the two excitations from both sides. The demonstrated all-dielectric absorbers pave a new path to realize a new class of absorbers with potential for novel applications, including imaging, energy harvesting, thermal emission, and sensors.

3.2.3   COHERENT PERFECT ABSORBERS

We have shown in Chapter 2 that under single-port incidence, only half of the incident power can be absorbed for a resonating system that supports only one mode at the resonant frequency.

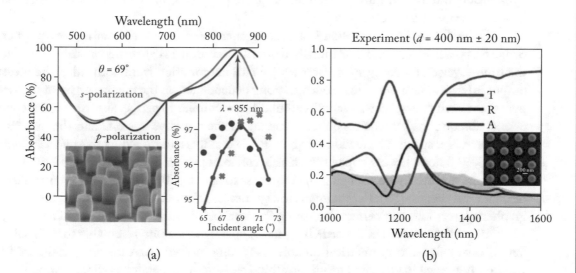

(a)   (b)

Figure 3.7: (a) Measured absorbance spectra $(1-R)$ of the silicon-nanopillar metasurface with height of 185-nm and diameter of 200-nm for p- and s-polarized light with incident angle 69° from the normal. The left inset shows a tilted SEM image of a typical metasurface consisting of silicon nanopillars. The right inset shows the angular dependence of the absorbance for $\lambda_0 = 855$-nm. The blue crosses (red circles) are for s- (p-) polarized incidence and the purple dotted line shows the mean absorption value. (b) Measured transmission, reflection, and absorption spectra for a NIR all-dielectric metasurface absorber with arrayed Ge disks on CaF_2 substrate. The height of disk is 160-nm, diameter is 400-nm, and the periodicity is 650-nm. The shadows denote the absorption spectra of a 160-nm-thin Ge film on the substrate. Reprinted with permission from: (a) Ref. [87] © 2017 ACS; (b) Ref. [88] © 2018 Wiley-VCH.

However, with an incoming wave from a second port, placed opposite to the first, perfect absorption may be possible if the two waves destructively interfere with the resonance in a system often termed a coherent perfect absorber (CPA). CPAs can be viewed as a time-reversed lasing process, where a gain medium is replaced with a lossy one [90]. Equivalently, the scattering matrix (S-matrix) of the CPA shown in Eq. (2.38) exhibits one zero, indicating decay of incident power, thus functioning as the counterpart to lasing. Absorption in such a system can be dynamically modulated through manipulating the phase, amplitude and frequency of the two incident waves—one at each port [91].

Coherent perfect absorption is a general phenomenon in a resonating cavity, when the S-matrix possesses zeros. It was initially demonstrated in a 110-μm silicon slab, which acts as the time-reverse of a Fabry-Pérot laser [92]. With illumination from both sides, the measured output intensity from one side revealed a strong dependence on the relative phase of the two inputs at the cavity resonant wavelengths around 1 μm. Similar to perfect absorption, the coherent perfect absorption also requires critical coupling of the radiative loss rate and the material loss rate. Therefore, spectral bandwidth of the CPA is mainly determined by the material loss tangent [93, 94]. For a heavily doped silicon slab, coherent perfect absorption in the Mid-IR range can also be achieved with changing the thickness to about 450-nm, which is about 1/20 of the resonant wavelength [95]. Recent numerical studies also showed that even for an atomically thin graphene layer, coherent perfect absorption is attainable via modifying the Fermi level [96, 97].

Recently, metamaterial based CPAs have generated great interest owing to their subwavelength cavity size and geometrical tunable resonating properties. A metamaterial-based CPA was first numerically presented on an ultrathin free-standing metamaterial film, which consists of two layers of metallic disk arrays separated by a dielectric spacer as shown in Fig. 3.8a [98]. Because of the plasmonic hybridization between the two disks, two fundamental modes, symmetric and anti-symmetric, correspond to the electric and magnetic dipole response, respectively. Both of these two modes are able to achieve coherent perfect absorption through varying the thickness of the dielectric spacer. For an optimized structure with a spacer thickness of $1/125\lambda_0$, where λ_0 is the resonant wavelength in vacuum, the absorption at the anti-symmetric mode shows a dependence on the relative phase between the two coherent beams, as shown in Fig. 3.8b. Around the same time, a metamaterial-based CPA was experimentally demonstrated at 632-nm using a free-standing metallic metamaterial with a thickness of $1/13\lambda_0$ [91]. Results illustrated the modulation of a signal intensity via manipulation of the phase or amplitude of a control beam of a Fano-type plasmonic mode. Due to the selective absorption on the coherency of photons, such a CPA has been used in quantum optics to demonstrate the entangled single photon states, which can be coupled to the plasmonic mode with nearly 100% probability [99].

However, not all of the coherent photons from the two ports can be directly absorbed into the resonator, but rather only those photons which match the phase (symmetry) of the resonating mode are absorbed. Through careful design of metamaterials with multiple modes of different symmetry, it is possible to achieve spectrally selective absorption of in-phase or out-of-

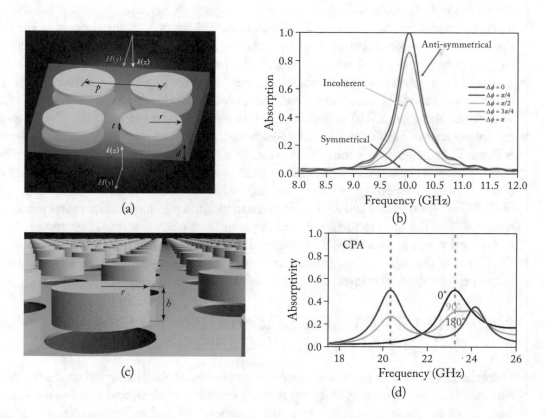

Figure 3.8: (a) Schematic of a plasmonic CPA with an MIM configuration. The top gold disk is separated from the bottom gold disk layer by a dielectric spacer. (b) Coherent two-port absorption of the microwave absorber with different phase difference. (c) Illustration of the all-dielectric CPA with arrayed zirconia cylinders. (d) Frequency-dependent two-port absorptivity with different phase difference from two ports for the designed CPA with diameter of 6.08 mm, and height of 1.4 mm. Reprinted with permission from: (a) and (b) Ref. [98] © 2012 NPG; (c) and (d) Ref. [48] © 2019 Wiley-VCH.

phase coherent photons [100]. Recently, such a scheme was first experimentally demonstrated using all-dielectric metasurfaces [48]. Figure 3.8c illustrates an all-dielectric CPA consisting of a square array of zirconia ceramic disks, which supports a large number of hybrid electric and magnetic resonances, including a quasi-electric dipole mode EH_{111}, and a quasi-magnetic dipole mode HE_{111}. According to dielectric waveguide theory and numerical simulations, for a given material loss tangent $\tan\delta = 0.06$ and permittivity of 31.6 for zirconia in the K-band, the disk array exhibits a perfect absorption state at 24.3 GHz, when the radius d and height h were set as 6.08 mm and 1.1 mm, respectively. At the perfect absorption frequency, the scattering matrix of the all-dielectric metasurface is zero because of the degeneracy of the critically coupled odd magnetic mode HE_{111}, and an even electric mode EH_{121}. Such an absorber is referred as a zero rank absorber (ZRA) as described in Chapter 2. As the height of the disks is increased, the frequencies of both modes decrease—although at different rates—such that the degeneracy is lifted. For a height of 1.4 mm, the two modes are well separated. Figure 3.8d shows the two-port absorptivity spectra for three different values of the port phase difference with $\Delta\phi = 0°$ (black curve), $\Delta\phi = 90°$ (grey curve), and $\Delta\phi = 180°$ (red curve). At the resonant frequencies of 20.3 GHz (odd mode) and 23.2 GHz (even mode), the phase dependent absorptivity reaches nearly 50% of the total incident incoherent power from one port. It is interesting to note that the critical coupling condition for both modes is approximately maintained while increasing the height of disks. Therefore, the spectrally selective CPA can be theoretically attained at any frequency in regions decoupled from high-order modes.

3.3   ELECTROMAGNETIC PROPERTIES OF METAMATERIAL ABSORBERS

Over the last decade since the emergence of metamaterial absorbers, potential practical applications in energy harvesting, imaging, thermal emission, sensing, and water vapor generation, have driven the exploration of metamaterial absorbers with various features, such as ultra-broad absorptivity bandwidth, narrow absorptivity bandwidth, polarization independent absorptivity, and spatially dependent absorptivity. In following sections, we will review the past achievements on these features and highlight a few particular examples.

3.3.1   BROAD BANDWIDTH

The first demonstrated MPA with a single unit cell exhibited only one absorption peak, and achieved relatively narrow bandwidth and high spectral resolution [1]. In single-resonator absorbers, the absorption has a relatively narrow bandwidth, and is achieved with only one mode, or two degenerated modes, with a finite loss rate. For broadband absorbers, the broader absorptivity operation spectrum means much lower quality factor (Q), and much larger material loss rate and radiative loss rate. Therefore, to achieve broadband absorptivity, methods with including lossy material or multiple resonators can be used to increase the loss rates, respectively.

To date, studies on broadband metamaterial perfect absorbers (BMPAs) can be categorized into four groups based on resonator arrangement and techniques: planar, vertical, integration with lumped elements, and randomly distributed nano-particles. The first straightforward method is to incorporate multiple resonators in a supercell through deliberate design with resonances overlapping each other. The final absorption spectrum, showing with multiple resonance peaks, can be approximately considered as the linear superposition of the individual resonance. Within a supercell consisting of a mix of resonators of various sizes, the resonators are close to each other such that the coupling between neighbors leads to interaction between each metamaterial, thus resulting in a slight resonance shift for each resonator. Dual-resonance MPAs were developed by two different groups at nearly the same time—both operating at terahertz frequencies [11, 70]. Both dual-band absorbers are designed with concatenating two ELC resonators together to form a single unit cell. As shown in Fig. 3.9a, due to the different resonator length, the absorber exhibits two strong absorption peaks of 0.85 at 1.41 THz and 0.94 at 3.02 THz, respectively. Using a similar method, with which resonators share the same unit cell in a single structure but have different either resonant length or gap size, dual-band [55, 101–108], triple-band [17, 109–115], quad-band [116–118], and octave-band [119, 120] metamaterial absorbers have been demonstrated in a spectrum, from microwave to visible.

Planar Broadband Absorbers

The first multispectral absorber demonstrated in the microwave region used two identical electrical circular ring resonators forming a 2 × 2 array in a unit cell by rotating two of the structures by 90°. The measured reflection shows two resonances at 11.15 and 16.01 GHz with absorptivity of 97% and 99%, respectively [122]. The multispectral absorption is achieved by matching the impedance of the structure with incorporating the fundamental ELC resonance and high-order dipole resonance mainly on the center bars. Because of the rotational symmetry, the absorptivity of such an absorber is polarization independent. With the same idea, other studies have shown multispectral absorbers in the microwave [123, 124], NIR [125–127], and at visible wavelengths [20]. As shown in Fig. 3.9d, an array of trapezoid shaped structure with Ag/SiO2/Ag layers reveals a broader absorptivity spectrum compared to a metallic grating array. A crossed trapezoidal pattern can further expand the absorption bandwidth, with absorptivity over 71% between 400 and 700-nm. The broadband absorption under TM-polarized excitation originates from the resonant electromagnetic response of different cross-sections of the trapezoid at different wavelength, while the absorption under TE polarization is due to the different effective plasma frequencies from different widths. Given the intrinsically weak absorption of ultrathin spacer layer, the absorbed power is all dissipated inside the top and bottom metal layers and subsequently converted to heat.

In the spectrum between microwave and NIR regimes, multispectral absorption can also be achieved by directly arranging resonators with various dimensions side by side within a supercell. As shown Fig. 3.9c, Liu et al., for the first time, achieved an NIR spectrally selective

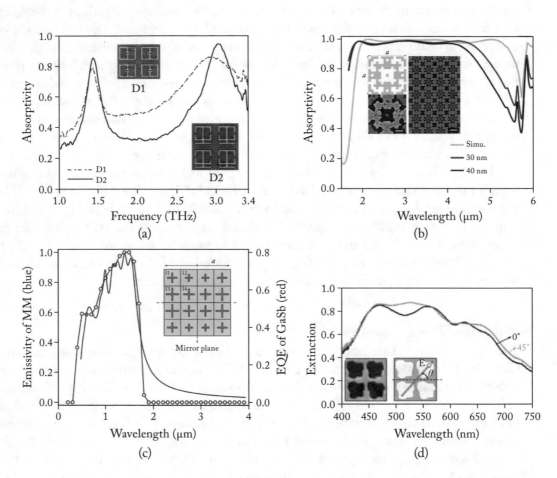

Figure 3.9: Planar arrangement of broadband MPAs. (a) Dual-band MPA utilizing single structure. Experimental absorptivity for single resonator (D1) and dual resonators (D2). (b) A broadband infrared absorber with super-octave bandwidth. Simulated and measured absorptivity spectra for Pd-based MPAs with Pd nanostructure thickness of 30 and 40-nm. The inset shows the SEM image of fabricated absorber with pattern generated via a genetic algorithm. (c) Broadband NIR emissivity spectra matching to the EQE of GaSb with 16 resonator sublattices. (d) A visible broadband metasurface absorber. Experimentally measured extinction spectra. The left inset shows an SEM image of the cross-trapezoid metamaterial and the right inset indicates the incident polarization angles. Reprinted with permission from: (a) Ref. [121] © 2010 IOP; (b) Ref. [119] © 2014 ACS; (c) Ref. [19] © 2011 APS; (d) Ref. [20] © 2011 NPG.

broadband absorber whose supercell consists of total 16 resonator elements with 4 different geometries [19]. The engineered emissivity spectrum operating in the range of 0.5–1.8 μm matches the external quantum efficiency (EQE) of GaSb to maximize the efficiency of energy harvesting using thermophotovoltaics (TPVs). Due to the subwavelength nature of the resonators, the total size of the supercell is only 1080 nm, smaller than the peak operating wavelength at 1.5 μm. The tessellation of unit cells can also be used for implementation of different geometries. Recently, an experimental demonstration of a broadband and polarization independent absorber, which consists of eight pairs of gold nano-resonators, including, cross, circular disks separated from a gold ground plane by a thin silicon dioxide layer [128]. The absorber exhibits absorptivity over 90% in the visible and near-infrared range of the solar spectrum. Similarly, such planar arrangements for broadband absorptivity have been demonstrated by many groups with operating center frequency from microwave to visible [128–141].

The planar arrangement method enables independent control of absorption bands in both magnitude and wavelength, thus providing flexibility to precisely engineer absorptivity or emissivity. The major shortcoming of this approach lies in that the operational bandwidth is limited by the number of resonators within a subwavelength supercell. In addition, the existence of many neighbors in a single super unit cell inevitably leads to interactions and spectral rearrangement. If neighboring interactions are significant, then achieving an accurate tailored scattering becomes difficult with conventional optimization approaches. To overcome the restrictions of planar arranged broadband metamaterial absorbers, optimization methods—such as genetic algorithms [119, 142] and adjoint optimization [143–145]—have been used to generated broadband absorbers with non-conventional geometries. Figure 3.9b shows an example of a genetic algorithm (GA) designed metamaterial array with polarization independent broadband absorptivity from 2–4.5 μm. The metamaterial was optimized by encoding one unit cell of the nano-structured array as a 15x15 binary pixels, and the geometry was forced to have 8-fold mirror symmetry to achieve polarization independent absorptivity. The algorithm evolves all of the adjustable design parameters with predefined ranges to identify the absorber with the target optical properties, including the thickness of each layer, periodicity, and patterned metal geometry features. The optimization process was performed via minimizing the cost function, which calculates the sum of deviation between the target optical properties and the simulated scattering properties over the target bandwidth and the range of incident angles for both polarizations. In recent years, taking advantage of the great interest from machine learning (ML) techniques, studies have shown the potential of ML for much more efficient metamaterial design [146–150].

Stacked Broadband Absorbers

Broadband metamaterial designs which use a planar arrangement are limited in both the fact that the filling fraction of any given sub-unit cell is reduced—thus reducing the oscillator strength of that mode [151]—as well as the fact that the super-unit cell size becomes comparable to the operational wavelength, which may lead to scattering. An alternative approach is to stack resonators vertically. In the vertical direction, more resonant layers can be added such that more absorption resonances are superposed with smaller intervals and, as a result, the operational bandwidth can be further expanded. The vertically stacked broadband absorber was first proposed by the He group operating at terahertz frequencies [152] and then demonstrated in other frequency regime [153–160]. As shown in Fig. 3.10a, the supercell consists of three layers of cross structure spaced with three layers of polymer and the dimensions of metamaterials gradually decrease from the bottom to the top layer. The top metamaterial with smallest geometry presents the highest resonant frequency because of the smallest wire length. The impedance-matching condition for each layer is achieved by tuning the polymer thickness such that the magnetic field at each resonance significantly concentrates between the corresponding layer and adjacent lower layer. The final total thickness of the stacked layers is still quite thin, smaller than $\lambda/15$. The numerical simulation shows three absorptivity peaks at 4.55 THz, 4.96 THz, and 5.37 THz. In this three-layer configuration, a 1.03 THz absorption band with absorptivity over 97% is achieved, and the FWHM of the absorption is nearly 38% of the central frequency.

To further expand and obtain a continuous high absorption spectrum, a sawtooth or truncated pyramid with tens of pairs of metal/dielectric stacked layers have been used [22, 159, 162–165]. The first vertically stacked ultrabroad absorbers was proposed in the mid-infrared regime by Fang and co-workers. Using a sawtooth structure with 20 pairs of metal/dielectric stacks as shown in Fig. 3.10c, the electromagnetic response can be described using effective anisotropic constitutive parameters $\varepsilon_{//}$ and $\varepsilon_{\perp}$, for the electric fields along directions parallel and perpendicular to metal/dielectric interface. Different from the stacked MPA with only a few layers, the incident wave propagates through the anisotropic metamaterial (AMM) in the z-direction and whirls into the AMM region, concentrating magnetic field close to the interface between AMM and air regions. As a result, the light acquires a very low group velocity compared to light speed in vacuum. Such an AMM absorber can be viewed as a group of air/AMM/air slowlight waveguides which can support slowlight modes with weakly coupled resonances from wavelength from 2–6 μm. An absorptivity over 90%, as shown in Fig. 3.10d, was obtained in the band from 3–6 μm. Such a high absorptivity can be maintained with incident angle up to 60°. The FWHM of the absorptivity spectrum is about 86% at normal incidence. An experimental demonstration of the stacked BMPA was achieved with 20 pairs of metal/dielectric stacked quadrangular frustum pyramid in X-band. The measured results showed over 90% absorptivity in the frequency range of 7.8–14.7 GHz, whose bandwidth is about 61% of the central frequency [163]. Additionally, the absorption is nearly independent on the incident angle below 40°, and remains above 80% with the incident angle within 60°.

A detailed investigation of the permittivity of stacked BMPAs revealed $\mathrm{Re}(\varepsilon_{||}) < 0$ and $\mathrm{Re}(\varepsilon_{\perp}) > 0$, indicating that the multilayered metamaterial is hyperbolic with an equifrequency dispersion curve hyperbolic [161, 166–170]. The hyperbolic property is strongly dependent on the thickness of dielectric layers. As the thickness increases, the value of the real part of the effective permittivity $\varepsilon_{\perp}$ turns from positive to negative. As a result, the hyperbolic property disap-

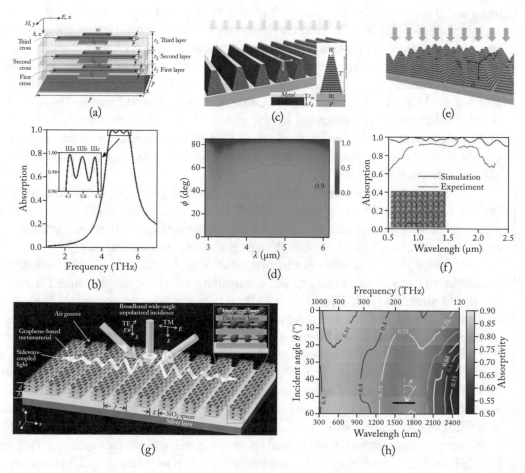

Figure 3.10: Vertical arrangement of BMPAs. (a), (b) First proposed vertically stacked broadband absorber. (a) Illustration of a three-layer stacked cross structure as a broadband absorber and its simulated absorption spectrum shown in (b). (c), (d) A sawtooth anisotropic infrared metamaterial with broadband absorptivity. (c) Illustration of the sawtooth structure with 20 pairs of Au/Ge stacks. (d) Simulated angular absorption spectrum and the line represents the absorptivity contour with value of 0.9. (e), (f) Experimental demonstration of a sawtooth hyperbolic BMPA. (e) Schematic of a hyperbolic BMPA consisting of two quadrangular frustum pyramids with different width. (f) Simulated (red solid line) and experimental measured (black dashed line) broadband absorptivity spectra. The inset shows an SEM image of the fabricated pyramids array. (g), (h) A broadband visible MPA using graphene/dielectric stacks. (g) Schematic of graphene-based BMPA. (b) Measured absorptivity versus wavelength and incident angle by an integrating sphere. Reprinted with permission from: (a) and (b) Ref. [152] 2010 © OSA; (c) and (d) Ref. [22] © 2012 ACS; (e) and (f) Ref. [161] © 2014 EMW Publishing; (g) and (h) Ref. [162] © 2019 NPG.

pears and the metal/dielectric stacks will no longer support a slow wave, leading to narrower absorption bandwidth [167]. To increase the absorption bandwidth, He and co-workers proposed and experimentally demonstrated an ultra-broad hyperbolic MPA by assembling multi-sized tapered patches to increase the absorption cross section, as shown in Fig. 3.10e. The hyperbolic MPA consisting of 10 pairs of alternating Al_2O_3 (16-nm)/Al (24-nm) thin layers on a 100-nm gold film was fabricated using a cross-beam system with focus-ion beam and scanning electron microscope. The measured results showed that the absorber yielded an average measured polarization-insensitive absorptivity of 85% in the entire solar spectrum, near-infrared and short-wavelength infrared regime from 500–2300 nm, with the bandwidth at FWHM over 128%, as shown in Fig.3.10f.

Recently, two-dimensional (2D) materials such as graphene, transition metal dichalcogenides (TMDs), and black phosphorus (BP) have garnered significant attention due to their great potential as tunable materials—enabled from strong light-matter interactions—for applications in photonics, optoelectronics, imaging, and telecommunications [171–173]. For vertically stacked BMPA, not only does incorporation of atomically thin 2D materials provide the possibility to dynamically modulate the absorptivity, via changing the carrier density or band gap of 2D materials [155, 174, 175], but also significantly decreases thermal mass of absorbers to further increase the speed of thermal response and temperature change [162], Lin et al. experimentally demonstrated a 90-nm-thick graphene MPA [162]. As shown in Fig. 3.10g, the BMPA consists of about 30 layers of graphene/dielectric stacks with a 80-nm SiO_2 spacer layer sitting on a 100-nm silver ground plane. The graphene/dielectric stack with a size as large as 25×50 mm^2 was fabricated using a wet chemical self-assembly technique, which utilizes static electric force between charged graphene oxide (GO) and polydiallyldimethylammonium chloride (PDDA) layers. The removal of oxygen-containing groups and recovery of graphene network was achieved using a femotosecond laser, which was also used for grating structure fabrication. Figure 3.10h shows the measured absorptivity versus wavelength and incident angle. Under normal incidence, the absorptivity is about 85% over the entire 300–2500-nm wavelength range. Further, characterization of the photothermal response showed that under natural sunlight in an open environment, the temperature of the absorber can increase by 130°C in 10 s.

Broadband Absorbers with Lumped Elements

At microwave and lower frequencies, planar and vertically arranged BMPA are relatively large and bulky. However, utilization of lumped element circuit components—resistors, capacitors, and inductors—can be directly integrated into the gaps of metamaterial unit cells, which not only modifies the effective impedance and loss, but also provides dynamical tuning capabilities [176–179]. Embedded lumped element capacitors and resistors in ELC and SRR resonators—combined together to form an absorber—showed a maximum absorption of 99.9% at 2.4 GHz with a FWHM of 700 MHz, which is about 29% of the central frequency [176].

Broadband Absorbers with Nanoparticles

In the near-infrared and visible wavelength ranges, fabrication of sub-microns features is more challenging and alternative methods of absorber manufacture may be more viable. For example, chemically and physically synthesized nanocomposites, which consist of nanoparticles with randomly distributed size, provide a new route to broadband absorption with the excitation of localized plasmon resonances. Through randomly dispersing chemically synthesized gold nanorod stacks on a gold film spaced by a dielectric layer, Chen et al. experimentally demonstrated a broadband absorber operating in the wavelength range of 900–1600-nm with absorption over 90% [180]. In addition to chemical synthesis, metal-dielectric nanocomposites can also be physically deposited using sputtering. Hedayati et al. showed that a layer of 20-nm thick $Ag-SiO_2$ nanocomposites can be co-sputtered onto a silver film with 10–30 nm SiO_2 spacer [30]. With randomly distributed sub-50-nm silver particles, the structure exhibited broadband near-unity absorption from deep UV region to 600 nm.

3.3.2   NARROW BANDWIDTH

High absorptivity with narrow bandwidth is crucial for spectroscopic chemical sensing, which requires discriminating chemical species with very close spectral fingerprints. Grating structures have been demonstrated with perfect absorption in the visible and infrared regions are enabled through coupling to surface plasmon polaritons [181–183] and surface-phonon polaritons [184]. Gratings consisting of arrayed patches [185], and nanoring/nanowire hybrid structures [34] on metal substrate without dielectric spacers, can achieve absorptivities exceeding 90% with FWHM as small as 0.4-nm in the near-infrared and 12-nm in the visible regions, respectively. However, since the absorption is based on momentum matching for the excitation of surface wave, the response is highly dependent on angle, with angular width on the order of mrad [185]. In contrast, the localized field excited in subwavelength metamaterials can significantly alleviate such limitations. According to temporal coupled-mode theory, absorption bandwidth is governed by the radiative loss rate of the resonant mode, and the material loss rate which depends on the material loss tangent. Narrowband perfect absorption therefore necessitates balancing a small radiative loss rate and material loss rate. Through tailoring the coupling between neighbouring units through increasing the length-to-width ratio and periodicity for a metallic cross-shaped resonator, one study experimentally showed that the ratio of FWHM to the operating wavelength can be reduced to about 3% as shown in Fig. 3.11 [186]. When two resonators with opposite mode symmetry are spaced close to each other, the coupling coefficient can be used to tweak the absorptivity. Recently, it was proposed that narrowband absorptivity of over 97% with a FWHM to wavelength ratio of 1.1% could be obtained through modification of the coupling phase between a dipole resonance and a quaduapole resonance [187].

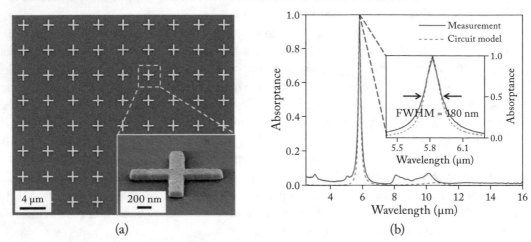

(a)   (b)

Figure 3.11: Narrowband IR absorber. (a) The false-colored SEM image of the ultra-narrow MIM IR absorber. The inset is a close-up SEM image of a cross-shaped nanostructure. (b) The measured absorption spectrum (black solid line) of the absorber compared with the absorption spectrum predicted by the equivalent circuit model (red dashed line). The inset highlights the ultra-narrow FWHM reaching its theoretical limit of 3%. Reprinted with permission from Ref. [186] Copyright 2019 Wiley-VCH.

3.3.3   POLARIZATION INDEPENDENT ABSORPTION

Resonances in metamaterials are—by construction—dependent on geometry, and thus the unit-cell symmetry determines the form of the magneto permittivity tensor and governs the resulting polarization dependence [69, 188–191]. To mitigate potential alignment issues of metamaterial with incident electric field, polarization independent absorptivity can be achieved by utilizing structures with 90° rotational symmetry [192–194]. The first demonstrated polarization independent absorber featured both top and bottom metallic structured with four-fold rotational symmetry through rotating primitive unit cell by 90° with respect to each other [1, 54]. Most of the demonstrated polarization-independent metamaterial absorber unit cells possess at least $\pi/2$ rotational symmetry [12, 153, 195]. However, it was shown that polarization independent absorptivity could be obtained with chiral metamaterials [196]. Rather than a standard three-layer absorber configuration, Soukoulis et al. presented a three-dimensional (3D) chiral structure, formed by two identical split-ring resonators separated by a dielectric substrate and interconnected by vias, providing electrical continuity for resonant currents. Unit cells are interlocked into a 3D grid sitting on a copper ground plane and covered with a dielectric plate. The large chirality enables both the electric and magnetic fields to induce strong resonances from which the incident power is dissipated inside the dielectric substrate.

3.3.4   WIDE ANGLE INCIDENCE

The angles over which high absorption is achieved, i.e., the acceptance angle, is critical for many applications. Generally, for metal-based MPAs with the canonical three-layer geometry, nearly all can absorb light to a relatively high angle of incidence in the microwave [16, 196–199], terahertz [6, 200, 201], infrared [57, 104, 131, 202–204], and visible [205]. In addition, most exhibit high absorptivity at large incident angles for both TE and TM polarizations. The first computational and experimental demonstration of wide-angle absorptivity verified that the absorptivity stays above 99% for TM polarization for the incidence within 80°, while absorption drops from 90% for incidence larger than 50° for TE polarization, due to the weak coupling to the magnetic field to drive circulating currents between the two metallic layers [6]. However, further investigation found that the wide-angle absorptivity for TM polarization originates from the coupling to a plasmon-like surface electromagnetic wave, which can be well described by the effective optical constants of the metamaterials [206].

3.4   DISCUSSION AND CONCLUSIONS

3.4.1   SUBWAVELENGTH GEOMETRY

A salient feature of metamaterials lies in that absorption peaks are primarily determined by the resonator geometry and the surrounding materials. In the long wavelength regime, such as microwave, millimeter wave, terahertz frequencies, and metamaterials can be easily obtained by winding metallic wires to obtain subwavelength resonant structures. Figure 3.12a–b list the most common geometries—electrical split-ring resonator (ESRR)—for terahertz and microwave absorbers, respectively [15, 21]. Chiral metamaterials, which exhibit cross coupling, i.e., chirality, between the electric and magnetic fields at resonance, have been used for demonstration of negative refraction. Through modification of the geometry, the circular polarized refractive indices ($n_\pm$) can be matched to free space, while most of the incident power is absorbed in the dielectric layer with total absorption over 98% for both circular polarizations, as shown in Fig. 3.12c [196]. Similar chiral metamaterial based absorbers were also demonstrated at infrared wavelengths [216]. Using advanced laser printing and electroplating techniques, more complicated structures with meander line or 3D shapes can be directly fabricated in the millimeter wave and terahertz ranges [201, 217]. Furthermore, at low frequencies, the relatively large metamaterial size enables incorporation of multiple structures and lumped elements into one unit cell providing multiple-band and broadband absorption [106, 177, 178]. At terahertz frequencies, the integration of functional materials, such as graphene, as shown in Fig. 3.12f [102] and GaAs [218], in the field-enhanced gap region of the resonator, permits large absorptive modulation depth.

For absorbers at infrared wavelengths, including long-wave, mid-wave, and near-infrared, direct down scaling of the SRRs operating at low frequencies becomes somewhat impractical due to limits in fabrication accuracy. Therefore, simple geometries, including dipole antennas [35,

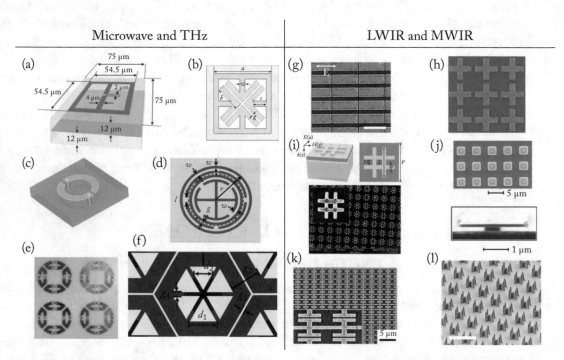

Figure 3.12: Typical metamaterial absorber geometries at frequencies from microwave to MWIR. (a) Electric split-ring resonator (ESRR) for terahertz absorbers. (b) A polarization-insensitive (PI) ESRR for a microwave absorber. (c) Chiral metamaterial unit cell for a microwave absorber. (d) A PI ESRR for bandwidth-enhanced microwave absorber. (e) A broadband MPA based integrated lumped elements. (f) A unit cell with six equilateral triangles in a regular hexagon with integrated graphene as a tunable THz absorber. (g) A dipole antenna based long-wave infrared (LWIR) absorber with integrated graphene for phase modulation. (h) Cross structure used for LWIR absorbers. (i) ♯ shape wires for an MWIR absorber. (j) A mushroom structure for IR absorbers. (k) I-beam metamaterial integrated with graphene for IR modulator. (l) 3D IR metamaterial absorbers. Reprinted with permission from: (a) Ref. [6] © 2008 APS; (b) Ref. [197] © 2011 AIP; (c) Ref. [196] © APS; (d) Ref. [106] © 2014 AIP; (e) Ref. [177] © 2012 AIP; (f) Ref. [207] © 2017 OSA; (g) Ref. [208] © 2017 ACS; (h) Ref. [12] © 2010 APS; (i) Ref. [104] © 2012 IOP; (j) Ref. [209] © 2015 AIP; (k) Ref. [71] © 2016 OSA; (l) Ref. [27] © 2013 Wiley-VCH.

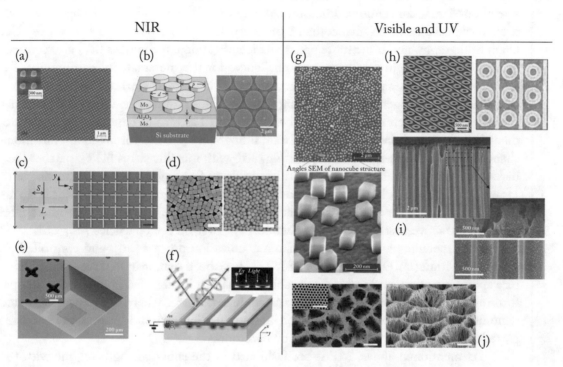

Figure 3.13: Typical metamaterial absorber geometries at frequencies from NIR to UV. (a) Dipole antenna for a NIR absorber. (b) Disk shape IR absorber. (c) A modified cross structural as a dual-band absorber. Scale bar = $2\,\mu$m. (d) Close-packed nanocubes and nanospheroids as NIR absorbers. Scale bar = 200 nm. (e) Babinet cross-based NIR absorber. (f) Illustration of a grating-based absorber for phase modulation. (g) Self-assembled nanocubes. (h) Nanoring and nanowire array for narrowband visible absorber. (i) Self-assembled nanoparticles for water desalination. (j) Self-aggregation of collapsed wires for water vapor generation. Reprinted with permission from: (a) Ref. [210] 2011 AIP; (b) Ref. [211] © 2016 Wiley-VCH; (c) Ref. [212] © 2012 ACS; (d) Ref. [213] © 2015 NPG; (e) Ref. [43] © 2016 NPG; (f) Ref. [214] © 2017 ACS; (g) Ref. [215] © 2017 Wiley-VCH; (h) Ref. [34] 2013 © ACS; (i) Ref. [36] © 2015 NPG; (j) Ref. [38] © 2016 NPG.

208, 219], crosses [12, 19, 119, 137, 140, 220, 221], square plates [39, 135], disks [133, 179, 211, 222–224], rings [225], holes [40], and I-beam shapes [71, 226], have been used for MPA demonstration as shown in Figs. 3.12 and 3.13. Matsuno and Sakurai demonstrated that when the critical dimension is larger than 2 μm for an arrayed disk metamaterial absorber with total size about 1 m^2, conventional photolithography can also used for patterning. However, for a desired critical dimension smaller than 1 μm, deep-UV lithography [12, 19, 227, 228], electron-beam lithography, and other more precise nanofabricational techniques [27] are required.

The visible regime possess more challenges for the metamaterial community. To accurately describe metamaterials with effective optical constants, $\varepsilon(\omega)$ and $\mu(\omega)$, the structures should appear homogeneous to incoming radiation. This requires the individual elements to be subwavelength, which requires element width as small as tens of nanometers. Such a requirement is very difficult even for electron beam lithography (EBL). Through simplification of the shapes of arrayed elements, as shown in Fig. 3.13 with metallic strips [111], patches [13, 138], disks [14, 125], or holes [229], it is still possible to fashion absorbers in the visible regime. However, the low fabricational yield and high cost of lithography methods for large area absorbers limit them for practical applications. Recently, large area visible absorbers fabricated by more attractive methods, such as chemical synthesized nanoparticles [23, 213, 230], self-aggregated metallic nanowire bundles from anodization process [36], and controlled physical vapor deposition (PVD) process [38, 231, 232] have been demonstrated for color filtering, water vapor generation, and water desalination. As shown in Figs. 3.13g, i–j, these nanoparticles forming nanocavities enable localized plasmonic resonances where light is concentrated on a nanometer-sized volume. Due to the non-uniform particle size and spacing, the self-assembled particles often result in a broad absorption spectrum.

As mentioned above, MPAs are fashioned in the subwavelength regime with the ratio of periodicity (p) to the wavelength λ_0 between 1:20 and 1:2 from microwave to the visible range [1, 6, 8, 12, 13, 19, 23, 56, 233, 234]. Through the design of a metamaterial structure as a spiral shape, it is possible for the ratio of p/λ_0 to reach 1:2000 [235]. In addition, MPAs provide an efficient means to reduce the device thickness, leading to a low profile. The original experimental design exhibited $\lambda_0/35$ and some groups have studied paths to achieve ultrathin absorbers [63, 236]. For example, absorbers have been experimentally demonstrated to be as thin as $\lambda_0/69$ and even $\lambda_0/75$ [197, 237]. As noticed, most of the metamaterials are designed in a unit-cell format with regular shapes. In this way, the near-field coupling between neighbors can be excluded. This design approach for a very subwavelength metamaterial is acceptable after comparing between simulations and experiments. However, for a complex metamaterial with various sub-units, the final performance cannot be simply added from those sub-units. The near-field coupling in a heterogeneous supercell will be more significant for all-dielectric metamaterial design [238]. With recent progress on the computational power and algorithms, novel design strategy, such as deep learning and inverse design methods could impact the metamaterial design disruptively, eventually bringing about unprecedented properties from irregular shapes.

Table 3.1: Materials used for constructing metamaterial in MMPA designs

	Metamaterial	References
Noble metals	Ag	[23, 127, 131, 239, 240]
	Al	[39, 179, 241, 242, 243]
	Au	[125, 244, 245, 246]
	Cu	[55, 234]
	Pt	[247]
Lossy transition metals	Cr	[248, 249]
	Mo	[211, 250]
	Ni	[251]
	Pd	[56, 126]
	Ti	[252, 251, 253]
	W	[41, 140, 142, 251, 254, 255]
	Eutectic GaIn	[256]
Conductive oxides and nitrides	ITO	[257]
	TiN	[46, 258]
Semiconductors	Graphene	[162]
	Silicon	[73, 74, 75, 76, 77]
	MoS_2	[259]

3.4.2   MATERIALS

Metals and dielectrics are the two major constituents used for constructing metal-dielectric-metal metamaterial absorber structures. Metals are crucial for resonating behavior and support electric responses, while the dielectric layer enables generation of a magnetic response between the two metal layers, while also providing a means to realize tunable absorbers. Tables 3.1 and 3.2 list the major conductive materials used for metamaterial layers, and dielectric materials for spacer layers shown in experimental demonstrations—only a few representative works are listed.

At low frequencies, e.g., the microwave regime, gold and copper are excellent conductors with very high conductivities, and are often used for making metamaterials. Aluminum and silver also possess high conductivities, but can be easily oxidized or corroded in air, and thus are more widely used at higher frequencies, i.e., infrared and visible wavelengths, because of their higher plasma frequencies. Aluminum and silver also have lower densities, which is attractive for potential application in satellites. All of the above-mentioned conductors are not transparent to visible light. To realize an optically transparent microwave absorber for microwave-proof windows, in place of a continuous metallic ground plane one may use an array of wires, mesh grid structures [271], or patterning the metamaterial and ground plane using transparent conductors such as indium tin oxide (ITO) [257], or aluminum-doped zinc oxide (AZO) [119].

Table 3.2: Dielectric materials as spacer for MMPA designs

Frequency	Dielectrics	References
Microwave and millimeter wave	FR4	[49, 234, 260, 261, 262]
	Rogers series	[24, 263]
	Water	[264, 265, 266, 267, 268, 269]
	Teflon	[270]
	Polyimide	[55]
	Carbon	[49]
	PDMS	[271]
	PMMA	[257]
Terahertz	Polyimide	[21, 25, 120, 218, 272]
	Kapton	[107]
	GaAs	[28, 33]
	SU-8	[273]
	Air	[274, 275]
	SiO_2	[223, 276]
	Ethanol	[277]
	Liquid crystal	[278]
Infrared	Al_2O_3	[13, 12, 19, 35, 45, 210, 279]
	HfO_2	[41, 280]
	CeO_2	[281]
	SiO_2	[125, 131, 186, 225, 250, 282]
	SiN_x	[208, 283, 284]
	MgF_2	[14, 141, 212, 285]
	a-Si	[226]
	Ge	[286, 287]
	ZnS	[49, 101, 129, 133]
	ZnO	[40]
	AlN	[288]
	$LiNbO_3$	[63]
	Kapton	[126]
	BARC*	[289]
	Air	[29, 43, 45, 229, 245]
	ITO	[46, 214, 280, 290]
	GST	[39, 224, 291]
Visible and UV	MgF_2	[56, 292]
	SiO_2	[20, 240, 242]
	PE layer**	[23, 215, 230]
	Al_2O_3	[64, 211, 247]
	TiO_2	[246, 293]
	a-Si	[239]

* bottom antireflection coating (Brewer Science: i-CON-7, i-CON-16)
** Poly(allylamine) hydrochloride (PAH) & Poly(styrenesulphonate) (PSS)

At microwave frequencies a common dielectric of choice for PCB substrates includes FR4 and the Rogers series, given their relative small loss tangent and abundant thickness choices. More recently, elastic polymers such as teflon [270], and polydimethylsiloxane (PDMS) [271], have been used to form flexible MMPAs at microwave frequencies. Metals identified above for the microwave regime still possess high conductivities at terahertz frequencies, and thus are often used. Polyimide is a good choice for a terahertz dielectric material, due to its small loss tangent, (≈ 0.04) ease of spin-coat fabrication [6, 8]. In addition to dielectric polymers, intrinsic, semi-insulating, or doped semiconductors, such as epitaxially grown GaAs, have been used for demonstrating terahertz metamaterials [28, 33]. An applied bias between the top and bottom metallic layers, enables tunable absorption between charge injection and depletion. Liquid crystals (LC) possess birefringence resulting in a δn ranging from 0.11–0.21, and are thus also viable candidates for reconfigurable terahertz absorbers [25, 278].

At the shorter wavelengths of infrared and optical, metals tend to be lossier and the Drude model is often used to describe their frequency dependent optical properties [66]. Selective emission and thermophotovoltaic (TPV) energy conversion in the infrared and visible ranges are important applications for metamaterial absorbers. Emitters/absorbers working at mid-infrared, near-infrared and visible range requires operational temperature higher than 1000°C, which is higher or close to the melting point of good conductors, such as Ag, Al, Cu, and Au. As a result, other transition metals with higher melting point are considered to be good candidates. Platinum (Pt) with a melting point of about 1800°C, and reasonably high conductivity, can be used for emitters operating in the mid-infrared range [247]. Tungsten (W) is more attractive for selective emission in the visible to NIR ranges due to its high melting point (3422°C) as well as low thermal expansion [41, 140, 142, 251, 254, 255]. However, tungsten is difficult to deposit and pattern, and thus refractory metals such as Cr, Mo, Ni, Ti, and TiN are also used for high temperature emission. Because of relative large imaginary part of permittivity, absorbers using refractory metals exhibit broad bandwidth in the NIR-visible range, presenting a compromise for high-temperature energy harvesting, such as solar TPV. In addition to use of conductive metals for MMPAs, semiconductors such as silicon and graphene possess a controllable and relatively high plasma frequency, thus enabling viable plasmonic properties for the construction of MMPAs.

Materials used as the dielectric spacer for operation in the visible-NIR are plentiful. CMOS compatible low-k dielectrics and semiconductors, such as SiO_2, SiN_x, MgF_2, Ge, α-Si, and germanium, can be directly coated through a PECVD or PVD process. However, the relatively low melting point and low quality of these materials is inapplicable for high temperature applications, such as SPTV. Recently, with the development of atomic layer deposition, high-quality dielectrics with high temperature melting points, as well as high dielectric constant, including Al_2O_3 [12, 13, 19, 35, 45, 64, 210, 211, 247, 279]. TiO_2 [246, 293], HfO_2 [41, 280], CeO_2 [281] have been widely used. These materials can be also implemented as protection layers coated on refractory metals from metal degradation in air [255, 294]. Phase change materials, like GST, VO$_2$, are dependent on the operational temperature with a phase transition between metallic state and insulating state. They are normally adopted for dynamic tuning of the absorptivity but have to worked at a specific temperature-controlled environment. ZnO, AlN, and LiNbO3 are piezoelectric materials, which are able to induce charge polarization and elasticity change from the changed temperature. Through directly replacing normal dielectric spacer with these functional piezoelectric materials between the metamaterial and ground plane, the optimized MPAs can be used for infrared detection [40, 40, 63]. At MWIR and NIR wavelengths, polymers with relatively small loss tangent, such as BARC (bottom antireflection coating) and Kapton can be spin coated on metals as a spacer with thickness sub-100-nm [126, 289]. To obtain ultrathin dielectric spacers for visible absorbers, 2-nm conformal coating of bilayer through directly immersing metal substrate into cationic and anionic solutions subsequently provides an efficient way for large area MPA fabrication [230].

3.4.3   CONCLUSION

The over one-decade endeavor on the metamaterial perfect absorbers has evolved the first experimental demonstration into an enormous forest with a variety of forms and characteristics. In this chapter, we introduced the metamaterial perfect absorbers via categorizing them based on the constitution materials, either metal or all dielectric. The characteristics of the perfect absorbers, such as the bandwidth, polarization dependence, angular dependence, have been described. Last, the frequently used materials and shapes for fashioning the metamaterial perfect absorbers are also generalized.

In addition to the absorbers we have mentioned, there are other approaches used to achieve perfect absorption, such as epsilon-near-zero (ENZ) media [295, 296], and gradient-metasurface based absorbers. Both absorbers exhibit the characteristics of converting the free-space propagating waves to bounded surface waves. In the ENZ media, the perfect absorption is achieved via critically coupling to the radiative Berreman mode and the non-radiative ENZ plasmon polariton mode [297–299]. Since these modes are TM polarization dependent, ENZ-type perfect absorbers can be used for ultrafast polarization switching through modifying the ENZ point of indium-doped cadmium oxide [300–302]. Alternatively, gradient-metasurfaces are able to guide incident waves to propagate along a dissipative surface through an extra momentum compensation from a space-variant phase gradient [303, 304]. It is also interesting that similar propagating-wave to surface-wave conversion was also observed on a conventional perfect absorber structure [305]. A TM-polarized incoming wave can couple to a plasmon-like surface mode, thus exhibiting a broad angular absorption response [305].

3.5   REFERENCES

[1] N. I. Landy, S. Sajuyigbe, J. J. Mock, D. R. Smith, and W. J. Padilla. Perfect metamaterial absorber. *Phys. Rev. Lett.*, 100(20):207402, May 2008. DOI: 10.1103/physrevlett.100.207402 29, 30, 32, 36, 44, 52, 56

[2] Winfield W Salisbury. Absorbent body for electromagnetic waves, Jun 1952. US Patent 2599944A. 29

[3] Ben A. Munk. *Frequency selective surfaces: theory and design.* John Wiley & Sons, Inc., April 2000. DOI: 10.1002/0471723770 29

[4] Xinyu Liu, Kebin Fan, Ilya V. Shadrivov, and Willie J. Padilla. Experimental realization of a terahertz all-dielectric metasurface absorber. *Opt. Express*, 25(1):191, Jan 2017. DOI: 10.1364/oe.25.000191 29, 32, 36, 38, 39

[5] Kebin Fan, Jonathan Y. Suen, Xinyu Liu, and Willie J. Padilla. All-dielectric metasurface absorbers for uncooled terahertz imaging. *Optica*, 4(6):601, Jun 2017. DOI: 10.1364/optica.4.000601 29, 32

[6] Hu Tao, C. M. Bingham, A. C. Strikwerda, D. Pilon, D. Shrekenhamer, N. I. Landy, K. Fan, X. Zhang, W. J. Padilla, and R. D. Averitt. Highly flexible wide angle of incidence terahertz metamaterial absorber: Design, fabrication, and characterization. *Phys. Rev. B*, 78(24):2–5, Dec 2008. DOI: 10.1103/physrevb.78.241103 29, 32, 33, 34, 53, 54, 56, 59

[7] Nader Engheta. Thin absorbing screens using metamaterial surfaces. In *IEEE Antennas and Propagation Society, AP-S International Symposium (Digest)*, volume 2, pages 392–395, 2002. DOI: 10.1109/aps.2002.1016106 32

[8] Hu Tao, Nathan I. Landy, Christopher M. Bingham, Xin Zhang, Richard D. Averitt, and Willie J. Padilla. A metamaterial absorber for the terahertz regime: design, fabrication and characterization. *Opt. Express*, 16(10):7181, May 2008. DOI: 10.1364/oe.16.007181 30, 32, 56, 59

[9] Jack Ng, Huanyang Chen, and C. T. Chan. Metamaterial frequency-selective superabsorber. *Opt. Lett.*, 34(5):644, Mar 2009. DOI: 10.1364/ol.34.000644 32

[10] Thomas Maier and Hubert Brückl. Wavelength-tunable microbolometers with metamaterial absorbers. *Opt. Lett.*, 34(19):3012, Oct 2009. DOI: 10.1364/ol.34.003012 32, 34

[11] Qi-Ye Wen, Huai-Wu Zhang, Yun-Song Xie, Qing-Hui Yang, and Ying-Li Liu. Dual band terahertz metamaterial absorber: Design, fabrication, and characterization. *Appl. Phys. Lett.*, 95(24):241111, Dec 2009. DOI: 10.1063/1.3276072 32, 45

[12] Xianliang Liu, Tatiana Starr, Anthony F. Starr, and Willie J. Padilla. Infrared spatial and frequency selective metamaterial with near-unity absorbance. *Phys. Rev. Lett.*, 104(20):207403, May 2010. DOI: 10.1103/physrevlett.104.207403 30, 32, 34, 35, 52, 54, 56, 59

[13] Jiaming Hao, Jing Wang, Xianliang Liu, Willie J. Padilla, Lei Zhou, and Min Qiu. High performance optical absorber based on a plasmonic metamaterial. *Appl. Phys. Lett.*, 96(25):251104, Jun 2010. DOI: 10.1063/1.3442904 32, 34, 35, 56, 59

[14] Na Liu, Martin Mesch, Thomas Weiss, Mario Hentschel, and Harald Giessen. Infrared perfect absorber and its application as plasmonic sensor. *Nano Lett.*, 10(7):2342–2348, Jul 2010. DOI: 10.1021/nl9041033 32, 34, 56

[15] Bo Zhu, Yijun Feng, Junming Zhao, Ci Huang, and Tian Jiang. Switchable metamaterial reflector/absorber for different polarized electromagnetic waves. *Appl. Phys. Lett.*, 97(5):051906, Aug 2010. DOI: 10.1063/1.3477960 32, 53

[16] Filiberto Bilotti, Alessandro Toscano, Kamil Boratay Alici, Ekmel Ozbay, and Lucio Vegni. Design of miniaturized narrowband absorbers based on resonant-magnetic inclusions. *IEEE Trans. Electromagn. Compat.*, 53(1):63–72, Feb 2011. DOI: 10.1109/temc.2010.2051229 32, 53

[17] Xiaopeng Shen, Tie Jun Cui, Junming Zhao, Hui Feng Ma, Wei Xiang Jiang, and Hui Li. Polarization-independent wide-angle triple-band metamaterial absorber. *Opt. Express*, 19(10):9401, May 2011. DOI: 10.1364/oe.19.009401 32, 45

[18] Yi Jin, Sanshui Xiao, N. Asger Mortensen, and Sailing He. Arbitrarily thin metamaterial structure for perfect absorption and giant magnification. *Opt. Express*, 19(12):11114, Jun 2011. DOI: 10.1364/oe.19.011114 32

[19] Xianliang Liu, Talmage Tyler, Tatiana Starr, Anthony F. Starr, Nan Marie Jokerst, and Willie J. Padilla. Taming the blackbody with infrared metamaterials as selective thermal emitters. *Phys. Rev. Lett.*, 107(4):045901, Jul 2011. DOI: 10.1103/physrevlett.107.045901 32, 36, 46, 47, 56, 59

[20] Koray Aydin, Vivian E. Ferry, Ryan M. Briggs, and Harry A. Atwater. Broadband polarization-independent resonant light absorption using ultrathin plasmonic super absorbers. *Nat. Commun.*, 2(1):517, Sep 2011. DOI: 10.1038/ncomms1528 32, 45, 46

[21] Krzysztof Iwaszczuk, Andrew C Strikwerda, Kebin Fan, Xin Zhang, Richard D Averitt, and Peter Uhd Jepsen. Flexible metamaterial absorbers for stealth applications at terahertz frequencies. *Opt. Express*, 20(1):635–43, Jan 2012. DOI: 10.1364/oe.20.000635 32, 53

[22] Yanxia Cui, Kin Hung Fung, Jun Xu, Hyungjin Ma, Yi Jin, Sailing He, and Nicholas X. Fang. Ultrabroadband light absorption by a sawtooth anisotropic metamaterial slab. *Nano Lett.*, 12(3):1443–1447, Mar 2012. DOI: 10.1021/nl204118h 32, 48, 49

[23] Antoine Moreau, Cristian Ciracì, Jack J. Mock, Ryan T. Hill, Qiang Wang, Benjamin J. Wiley, Ashutosh Chilkoti, and David R. Smith. Controlled-reflectance surfaces with film-coupled colloidal nanoantennas. *Nature*, 492(7427):86–89, Dec 2012. DOI: 10.1038/nature11615 32, 56

[24] David Shrekenhamer, Wangren Xu, Suresh Venkatesh, David Schurig, Sameer Sonkusale, and Willie J. Padilla. Experimental realization of a metamaterial detector focal plane array. *Phys. Rev. Lett.*, 109(17):177401, Oct 2012. DOI: 10.1103/physrevlett.109.177401 32

[25] David Shrekenhamer, Wen-Chen Chen, and Willie J. Padilla. Liquid crystal tunable metamaterial absorber. *Phys. Rev. Lett.*, 110(17):177403, Apr 2013. DOI: 10.1103/physrevlett.110.177403 32, 59

[26] Fabio Alves, Dragoslav Grbovic, Brian Kearney, Nickolay V. Lavrik, and Gamani Karunasiri. Bi-material terahertz sensors using metamaterial structures. *Opt. Express*, 21(11):13256, May 2013. DOI: 10.1364/oe.21.013256 32

[27] Xiang Xiong, Shang-Chi Jiang, Yu-Hui Hu, Ru-Wen Peng, and Mu Wang. Structured Metal Film as a Perfect Absorber. *Adv. Mater.*, 25(29):3994–4000, Aug 2013. DOI: 10.1002/adma.201300223 32, 54, 56

[28] David Shrekenhamer, John Montoya, Sanjay Krishna, and Willie J. Padilla. Four-color metamaterial absorber THz spatial light modulator. *Adv. Opt. Mater.*, 1(12):905–909, Dec 2013. DOI: 10.1002/adom.201300265 32, 59

[29] Xianliang Liu and Willie J. Padilla. Dynamic manipulation of infrared radiation with MEMS metamaterials. *Adv. Opt. Mater.*, pages n/a–n/a, Jun 2013. DOI: 10.1002/adom.201300163 32, 36

[30] M. K. Hedayati, A. U. Zillohu, T. Strunskus, F. Faupel, and M. Elbahri. Plasmonic tunable metamaterial absorber as ultraviolet protection film. *Appl. Phys. Lett.*, 104(4):041103, Jan 2014. DOI: 10.1063/1.4863202 32, 51

[31] Fusheng Ma, Yu-Sheng Lin, Xinhai Zhang, and Chengkuo Lee. Tunable multiband terahertz metamaterials using a reconfigurable electric split-ring resonator array. *Light Sci. Appl.*, 3(5):e171–e171, May 2014. DOI: 10.1038/lsa.2014.52 32

[32] Wei Li and Jason Valentine. Metamaterial perfect absorber based hot electron photodetection. *Nano Lett.*, 14(6):3510–3514, Jun 2014. DOI: 10.1021/nl501090w 32

[33] Claire M. Watts, David Shrekenhamer, John Montoya, Guy Lipworth, John Hunt, Timothy Sleasman, Sanjay Krishna, David R. Smith, and Willie J. Padilla. Terahertz compressive imaging with metamaterial spatial light modulators. *Nat. Photonics*, 8(8):605–609, Jun 2014. DOI: 10.1038/nphoton.2014.139 32, 59

[34] Zhongyang Li, Serkan Butun, and Koray Aydin. Ultranarrow Band Absorbers Based on Surface Lattice Resonances in Nanostructured Metal Surfaces. *ACS Nano*, 8(8):8242–8248, Aug 2014. DOI: 10.1021/nn502617t 32, 51, 55

[35] Yu Yao, Raji Shankar, Mikhail A Kats, Yi Song, Jing Kong, Marko Loncar, and Federico Capasso. Electrically tunable metasurface perfect absorbers for ultrathin mid-infrared optical modulators. *Nano Lett.*, 14(11):6526–32, Nov 2014. DOI: 10.1021/nl503104n 32, 53, 59

[36] Kyuyoung Bae, Gumin Kang, Suehyun K. Cho, Wounjhang Park, Kyoungsik Kim, and Willie J. Padilla. Flexible thin-film black gold membranes with ultrabroadband plasmonic nanofocusing for efficient solar vapour generation. *Nat. Commun.*, 6:10103, Dec 2015. DOI: 10.1038/ncomms10103 32, 34, 55, 56

[37] Zhongyang Li, Serkan Butun, and Koray Aydin. Large-area, lithography-free super absorbers and color filters at visible frequencies using ultrathin metallic films. *ACS Photonics*, 2(2):183–188, Feb 2015. DOI: 10.1021/ph500410u 32

[38] Lin Zhou, Yingling Tan, Jingyang Wang, Weichao Xu, Ye Yuan, Wenshan Cai, Shining Zhu, and Jia Zhu. 3D self-assembly of aluminium nanoparticles for plasmon-enhanced solar desalination. *Nat. Photonics*, 10(6):393–398, Apr 2016. DOI: 10.1038/nphoton.2016.75 32, 34, 55, 56

[39] Andreas Tittl, Ann-Katrin U. Michel, Martin Schäferling, Xinghui Yin, Behrad Gholipour, Long Cui, Matthias Wuttig, Thomas Taubner, Frank Neubrech, and Harald Giessen. A switchable mid-infrared plasmonic perfect absorber with multispectral thermal imaging capability. *Adv. Mater.*, 27(31):4597–4603, Aug 2015. DOI: 10.1002/adma.201502023 32, 56

[40] Thang Duy Dao, Satoshi Ishii, Takahiro Yokoyama, Tomomi Sawada, Ramu Pasupathi Sugavaneshwar, Kai Chen, Yoshiki Wada, Toshihide Nabatame, and Tadaaki Nagao. Hole array perfect absorbers for spectrally selective midwavelength infrared pyroelectric detectors. *ACS Photonics*, 3(7):1271–1278, Jul 2016. DOI: 10.1021/acsphotonics.6b00249 32, 34, 56, 60

[41] P. N. Dyachenko, S. Molesky, A. Yu Petrov, M. Störmer, T. Krekeler, S. Lang, M. Ritter, Z. Jacob, and M. Eich. Controlling thermal emission with refractory epsilon-near-zero metamaterials via topological transitions. *Nat. Commun.*, 7(1):11809, Dec 2016. DOI: 10.1038/ncomms11809 32, 59

[42] Yao-Wei Huang, Ho Wai Howard Lee, Ruzan Sokhoyan, Ragip A. Pala, Krishnan ThyagaraJan, Seunghoon Han, Din Ping Tsai, and Harry A. Atwater. Gate-tunable conducting oxide metasurfaces. *Nano Lett.*, 16(9):5319–5325, Sep 2016. DOI: 10.1021/acs.nanolett.6b00555 32

[43] Hai Zhu, Fei Yi, and Ertugrul Cubukcu. Plasmonic metamaterial absorber for broadband manipulation of mechanical resonances. *Nat. Photonics*, 10(11):709–714, Nov 2016. DOI: 10.1038/nphoton.2016.183 32, 55

[44] Yao Zhai, Ma Yaoguang, Sabrina N. David, Dongliang Zhao, Runnan Lou, Gang Tan, Ronggui Yang, and Xiaobo Yin. Scalable-manufactured randomized glass-polymer hybrid metamaterial for daytime radiative cooling. *Science*, 355(6329):1062–1066, Mar 2017. DOI: 10.1126/science.aai7899 32

[45] Xinyu Liu and Willie J. Padilla. Reconfigurable room temperature metamaterial infrared emitter. *Optica*, 4(4):430, Apr 2017. DOI: 10.1364/optica.4.000430 32, 59

[46] Joshua R. Hendrickson, Shivashankar Vangala, Chandriker Dass, Ricky Gibson, John Goldsmith, Kevin Leedy, Dennis E. Walker, Justin W. Cleary, Wonkyu Kim, and Junpeng Guo. Coupling of epsilon-near-zero mode to gap plasmon mode for flat-top wideband perfect light absorption. *ACS Photonics*, 5(3):776–781, Mar 2018. DOI: 10.1021/acsphotonics.7b01491 32

[47] Kebin Fan, Jingdi Zhang, Xinyu Liu, Gu-Feng Zhang, Richard D. Averitt, and Willie J. Padilla. Phototunable dielectric Huygens' metasurfaces. *Adv. Mater.*, 30(22):1800278, May 2018. DOI: 10.1002/adma.201800278 32, 38

[48] J. Y. Suen, K. Fan, and W. J. Padilla. A zero-rank, maximum nullity perfect electromagnetic wave absorber. *Adv. Opt. Mater.*, page 1801632, Jan 2019. DOI: 10.1002/adom.201801632 32, 40, 43, 44

[49] Taehwan Kim, Ji-Yeul Bae, Namkyu Lee, and Hyung Hee Cho. Hierarchical metamaterials for multispectral camouflage of infrared and microwaves. *Adv. Funct. Mater.*, 29(10):1807319, Mar 2019. DOI: 10.1002/adfm.201807319 32

[50] Kevin Pichler, Matthias Kühmayer, Julian Böhm, Andre Brandstötter, Philipp Ambichl, Ulrich Kuhl, and Stefan Rotter. Random anti-lasing through coherent perfect absorption in a disordered medium. *Nature*, 567(7748):351–355, Mar 2019. DOI: 10.1038/s41586-019-0971-3 32

[51] Mohammadreza F. Imani, David R. Smith, and Philipp Hougne. Perfect absorption in a disordered medium with programmable meta-atom inclusions. *Advanced Functional Materials*, 30(52):2005310, Dec 2020. DOI: 10.1002/adfm.202005310 32

[52] Léo Wojszvzyk, Anne Nguyen, Anne-Lise Coutrot, Cheng Zhang, Benjamin Vest, and Jean-Jacques Greffet. An incandescent metasurface for quasimonochromatic polarized mid-wave infrared emission modulated beyond 10 MHz. *Nature Communications*, 12(1), March 2021. DOI: 10.1038/s41467-021-21752-w 32

[53] Taowu Deng, Jiangang Liang, Tong Cai, Canyu Wang, Xin Wang, Jing Lou, Zhiqiang Du, and Dengpan Wang. Ultra-thin and broadband surface wave meta-absorber. *Optics Express*, 29(12):19193, June 2021. DOI: 10.1364/oe.427992 32

[54] N. I. Landy, C. M. Bingham, T. Tyler, N. Jokerst, D. R. Smith, and W. J. Padilla. Design, theory, and measurement of a polarization-insensitive absorber for terahertz imaging. *Phys. Rev. B*, 79(12):125104, Mar 2009. DOI: 10.1103/physrevb.79.125104 30, 34, 52

[55] Pramod K. Singh, Konstantin A. Korolev, Mohammed N. Afsar, and Sameer Sonkusale. Single and dual band 77/95/110 GHz metamaterial absorbers on flexible polyimide substrate. *Appl. Phys. Lett.*, 99(26):264101, Dec 2011. DOI: 10.1063/1.3672100 34, 35, 45

[56] Andreas Tittl, Patrick Mai, Richard Taubert, Daniel Dregely, Na Liu, and Harald Giessen. Palladium-based plasmonic perfect absorber in the visible wavelength range and its application to hydrogen sensing. *Nano Lett.*, 11(10):4366–4369, Oct 2011. DOI: 10.1021/nl202489g 35, 36, 56

[57] Yoav Avitzour, Yaroslav A. Urzhumov, and Gennady Shvets. Wide-angle infrared absorber based on a negative-index plasmonic metamaterial. *Phys. Rev. B*, 79(4):045131, Jan 2009. DOI: 10.1103/physrevb.79.045131 34, 53

[58] S. A. Kuznetsov, A. G. Paulish, A. V. Gelfand, P. A. Lazorskiy, and V. N. Fedorinin. Bolometric THz-to-IR converter for terahertz imaging. *Appl. Phys. Lett.*, 99(2):023501, 2011. DOI: 10.1063/1.3607474 34

[59] Fabio Alves, Brian Kearney, Dragoslav Grbovic, Nickolay V. Lavrik, and Gamani Karunasiri. Strong terahertz absorption using SiO_2 /Al based metamaterial structures. *Appl. Phys. Lett.*, 100(11):111104, Mar 2012. DOI: 10.1063/1.3693407 34

[60] James Grant, Ivonne Escorcia-Carranza, Chong Li, Iain J. H. McCrindle, John Gough, and David R. S. Cumming. A monolithic resonant terahertz sensor element comprising a metamaterial absorber and micro-bolometer. *Laser Photon. Rev.*, 7(6):1043–1048, Nov 2013. DOI: 10.1002/lpor.201300087 34

[61] John Lehman, Evangelos Theocharous, George Eppeldauer, and Chris Pannell. Gold-black coatings for freestanding pyroelectric detectors. *Meas. Sci. Technol.*, 14(7):916–922, Jul 2003. DOI: 10.1088/0957-0233/14/7/304 34

[62] John H. Lehman, Chaiwat Engtrakul, Thomas Gennett, and Anne C. Dillon. Single-wall carbon nanotube coating on a pyroelectric detector. *Appl. Opt.*, 44(4):483, Feb 2005. DOI: 10.1364/ao.44.000483 34

[63] J. Y. Suen, K. Fan, J. Montoya, C. Bingham, Vincent Stenger, S. Sriram, and W. J. Padilla. Multifunctional metamaterial pyroelectric infrared detectors. *Optica*, 4(2):276, Feb 2017. DOI: 10.1364/optica.4.000276 34, 56, 60

[64] Jing Wang, Yiting Chen, Xi Chen, Jiaming Hao, Min Yan, and Min Qiu. Photothermal reshaping of gold nanoparticles in a plasmonic absorber. *Opt. Express*, 19(15):14726, Jul 2011. DOI: 10.1364/oe.19.014726 34, 59

[65] Jiaming Hao, Lei Zhou, and Min Qiu. Nearly total absorption of light and heat generation by plasmonic metamaterials. *Phys. Rev. B*, 83(16):165107, Apr 2011. DOI: 10.1103/physrevb.83.165107 34

[66] M A Ordal, Robert J Bell, Jr R W Alexander, L L Long, and M R Querry. Optical properties of fourteen metals in the infrared and far infrared: Al, Co, Cu, Au, Fe, Pb, Mo, Ni, Pd, Pt, Ag, Ti, V, and W. *Appl. Opt.*, 24(24):4493–4499, Dec 1985. DOI: 10.1364/ao.24.004493 36, 59

[67] Stefan A. Maier. *Plasmonics: Fundamentals and Applications*. Springer US, New York, NY, 2007. DOI: 10.1007/0-387-37825-1 36

[68] J. B. Pendry, A. J. Holden, W. J. Stewart, and I. Youngs. Extremely low frequency plasmons in metallic mesostructures. *Phys. Rev. Lett.*, 76(25):4773–4776, Jun 1996. DOI: 10.1103/physrevlett.76.4773 36

[69] J.B. Pendry, A.J. Holden, D.J. Robbins, and W.J. Stewart. Magnetism from conductors and enhanced nonlinear phenomena. *IEEE Trans. Microw. Theory Tech.*, 47(11):2075–2084, 1999. DOI: 10.1109/22.798002 36, 52

[70] Hu Tao, Jason J. Amsden, Andrew C. Strikwerda, Kebin Fan, David L. Kaplan, Xin Zhang, Richard D. Averitt, and Fiorenzo G. Omenetto. Metamaterial silk composites at terahertz frequencies. *Advanced Materials*, 22(32):3527–3531, Aug 2010. DOI: 10.1002/adma.201000412 36, 45

[71] Kebin Fan, Jonathan Suen, Xueyuan Wu, and Willie J. Padilla. Graphene metamaterial modulator for free-space thermal radiation. *Optics Express*, 24(22):25189, Oct 2016. DOI: 10.1364/oe.24.025189 36, 54, 56

[72] Christian C. Nadell, Claire M. Watts, John A. Montoya, Sanjay Krishna, and Willie J. Padilla. Single pixel quadrature imaging with metamaterials. *Adv. Opt. Mater.*, 4(1):66–69, Jan 2016. DOI: 10.1002/adom.201500435 36

[73] Mingbo Pu, Min Wang, Chenggang Hu, Cheng Huang, Zeyu Zhao, Yanqin Wang, and Xiangang Luo. Engineering heavily doped silicon for broadband absorber in the terahertz regime. *Opt. Express*, 20(23):25513, Nov 2012. DOI: 10.1364/oe.20.025513 37, 38

[74] Yong Zhi Cheng, Withawat Withayachumnankul, Aditi Upadhyay, Daniel Headland, Yan Nie, Rong Zhou Gong, Madhu Bhaskaran, Sharath Sriram, and Derek Abbott. Ultrabroadband plasmonic absorber for terahertz waves. *Adv. Opt. Mater.*, 3(3):376–380, Mar 2015. DOI: 10.1002/adom.201400368 37, 38

[75] Withawat Withayachumnankul, Charan Manish Shah, Christophe Fumeaux, Benjamin S.-Y. Ung, Willie J. Padilla, Madhu Bhaskaran, Derek Abbott, and Sharath Sriram. Plasmonic resonance toward terahertz perfect absorbers. *ACS Photonics*, 1(7):625–630, Jul 2014. DOI: 10.1021/ph500110t 37, 38

[76] Cheng Shi, XiaoFei Zang, YiQiao Wang, Lin Chen, Bin Cai, and YiMing Zhu. A polarization-independent broadband terahertz absorber. *Appl. Phys. Lett.*, 105(3):031104, Jul 2014. DOI: 10.1063/1.4890617 37

[77] Sheng Yin, Jianfei Zhu, Wendao Xu, Wei Jiang, Jun Yuan, Ge Yin, Lijuan Xie, Yibin Ying, and Yungui Ma. High-performance terahertz wave absorbers made of silicon-based metamaterials. *Appl. Phys. Lett.*, 107(7):073903, Aug 2015. DOI: 10.1063/1.4929151 37

[78] Kazim Gorgulu, Abdullah Gok, Mehmet Yilmaz, Kagan Topalli, Necmi Bıyıklı, and Ali K. Okyay. All-silicon ultra-broadband infrared light absorbers. *Sci. Rep.*, 6(1):38589, Dec 2016. DOI: 10.1038/srep38589 37

[79] Manuel Decker, Isabelle Staude, Matthias Falkner, Jason Dominguez, Dragomir N. Neshev, Igal Brener, Thomas Pertsch, and Yuri S. Kivshar. High-efficiency dielectric Huygens' surfaces. *Adv. Opt. Mater.*, 3(6):813–820, Jun 2015. DOI: 10.1002/adom.201400584 38

[80] Zhijie Ma, Stephen M. Hanham, Pablo Albella, Binghao Ng, Hsiao Tzu Lu, Yandong Gong, Stefan A. Maier, and Minghui Hong. Terahertz all-dielectric magnetic mirror metasurfaces. *ACS Photonics*, 3(6):1010–1018, Jun 2016. DOI: 10.1021/acsphotonics.6b00096 38

[81] Daniel Headland, Eduardo Carrasco, Shruti Nirantar, Withawat Withayachumnankul, Philipp Gutruf, James Schwarz, Derek Abbott, Madhu Bhaskaran, Sharath Sriram, Julien Perruisseau-Carrier, and Christophe Fumeaux. Dielectric resonator reflectarray as high-efficiency nonuniform terahertz metasurface. *ACS Photonics*, 3(6):1019–1026, Jun 2016. DOI: 10.1021/acsphotonics.6b00102 38

[82] Isabelle Staude, Andrey E. Miroshnichenko, Manuel Decker, Nche T. Fofang, Sheng Liu, Edward Gonzales, Jason Dominguez, Ting Shan Luk, Dragomir N. Neshev, Igal Brener, and Yuri Kivshar. Tailoring directional scattering through magnetic and electric resonances in subwavelength silicon nanodisks. *ACS Nano*, 7(9):7824–7832, Sep 2013. DOI: 10.1021/nn402736f 38

[83] Jürgen Sautter, Isabelle Staude, Manuel Decker, Evgenia Rusak, Dragomir N. Neshev, Igal Brener, and Yuri S. Kivshar. Active tuning of all-dielectric metasurfaces. *ACS Nano*, 9(4):4308–4315, Apr 2015. DOI: 10.1021/acsnano.5b00723 38

[84] Katie E. Chong, Lei Wang, Isabelle Staude, Anthony R. James, Jason Dominguez, Sheng Liu, Ganapathi S. Subramania, Manuel Decker, Dragomir N. Neshev, Igal Brener, and Yuri S. Kivshar. Efficient polarization-insensitive complex wavefront control using Huygens' metasurfaces based on dielectric resonant meta-atoms. *ACS Photonics*, 3(4):514–519, Apr 2016. DOI: 10.1021/acsphotonics.5b00678 38

[85] Michael A Cole, David A Powell, and Ilya V Shadrivov. Strong terahertz absorption in all-dielectric Huygens' metasurfaces. *Nanotechnology*, 27(42):424003, Oct 2016. DOI: 10.1088/0957-4484/27/42/424003 38

[86] Rajesh K. Mongia and Prakash Bhartia. Dielectric resonator antennas—a review and general design relations for resonant frequency and bandwidth. *International Journal of Microwave and Millimeter-Wave Computer-Aided Engineering*, 4(3):230–247, Jul 1994. DOI: 10.1002/mmce.4570040304 38

[87] Nils Odebo Länk, Ruggero Verre, Peter Johansson, and Mikael Käll. Large-scale silicon nanophotonic metasurfaces with polarization independent near-perfect absorption. *Nano Lett.*, 17(5):3054–3060, May 2017. DOI: 10.1021/acs.nanolett.7b00416 40, 41

[88] Jingyi Tian, Hao Luo, Qiang Li, Xuelu Pei, Kaikai Du, and Min Qiu. Near-infrared super-absorbing all-dielectric metasurface based on single-layer germanium nanostructures. *Laser Photon. Rev.*, 12(9):1800076, Sep 2018. DOI: 10.1002/lpor.201800076 40, 41

[89] Chi-Yin Yang, Jhen-Hong Yang, Zih-Ying Yang, Zhong-Xing Zhou, Mao-Guo Sun, Viktoriia E. Babicheva, and Kuo-Ping Chen. Nonradiating silicon nanoantenna metasurfaces as narrowband absorbers. *ACS Photonics*, 5(7):2596–2601, Jul 2018. DOI: 10.1021/acsphotonics.7b01186 40

[90] Y. D. Chong, Li Ge, Hui Cao, and A. D. Stone. Coherent perfect absorbers: time-reversed lasers. *Physical Review Letters*, 105(5):053901, Jul 2010. DOI: 10.1103/physrevlett.105.053901 42

[91] Jianfa Zhang, Kevin F. MacDonald, and Nikolay I. Zheludev. Controlling light-with-light without nonlinearity. *Light: Science and Applications*, 1(JulY):e18–e18, Jul 2012. DOI: 10.1038/lsa.2012.18 42

[92] Wenjie Wan, Yidong Chong, Li Ge, Heeso Noh, A Douglas Stone, and Hui Cao. Time-reversed lasing and interferometric control of absorption. *Science (New York, N.Y.)*, 331(6019):889–92, Feb 2011. DOI: 10.1126/science.1200735 42

[93] Simin Feng and Klaus Halterman.   Coherent perfect absorption in epsilon-near-zero metamaterials.   *Phys. Rev. B*, 86(16):165103, Oct 2012. DOI: 10.1103/phys-revb.86.165103 42

[94] Sucheng Li, Jie Luo, Shahzad Anwar, Shuo Li, Weixin Lu, Zhi Hong Hang, Yun Lai, Bo Hou, Mingrong Shen, and Chinhua Wang. Broadband perfect absorption of ultrathin conductive films with coherent illumination: Superabsorption of microwave radiation. *Phys. Rev. B*, 91:220301, Jun 2015. DOI: 10.1103/physrevb.91.220301 42

[95] Mingbo Pu, Qin Feng, Min Wang, Chenggang Hu, Cheng Huang, Xiaoliang Ma, Zeyu Zhao, Changtao Wang, and Xiangang Luo.   Ultrathin broadband nearly perfect absorber with symmetrical coherent illumination. *Opt. Express*, 20(3):2246, Jan 2012. DOI: 10.1364/oe.20.002246 42

[96] Jianfa Zhang, Chucai Guo, Ken Liu, Zhihong Zhu, Weimin Ye, Xiaodong Yuan, and Shiqiao Qin. Coherent perfect absorption and transparency in a nanostructured graphene film. *Opt. Express*, 22(10):12524, May 2014. DOI: 10.1364/oe.22.012524 42

[97] Yuancheng Fan, Zhe Liu, Fuli Zhang, Qian Zhao, Zeyong Wei, Quanhong Fu, Junjie Li, Changzhi Gu, and Hongqiang Li. Tunable mid-infrared coherent perfect absorption in a graphene meta-surface. *Sci. Rep.*, 5(1):13956, Nov 2015. DOI: 10.1038/srep13956 42

[98] Mingbo Pu, Qin Feng, Chenggang Hu, and Xiangang Luo. Perfect absorption of light by coherently induced plasmon hybridization in ultrathin metamaterial film. *Plasmonics*, 7(4):733–738, April 2012. DOI: 10.1007/s11468-012-9365-1 42, 43

[99] Thomas Roger, Stefano Vezzoli, Eliot Bolduc, Joao Valente, Julius J. F. Heitz, John Jeffers, Cesare Soci, Jonathan Leach, Christophe Couteau, Nikolay I. Zheludev, and Daniele Faccio. Coherent perfect absorption in deeply subwavelength films in the single-photon regime. *Nat. Commun.*, 6(1):7031, Dec 2015. DOI: 10.1038/ncomms8031 42

[100] Guangyu Nie, Quanchao Shi, Zheng Zhu, and Jinhui Shi.   Selective coherent perfect absorption in metamaterials. *Appl. Phys. Lett.*, 105(20):201909, Nov 2014. DOI: 10.1063/1.4902330 44

[101] Charlie Koechlin, Patrick Bouchon, Fabrice Pardo, Julien Jaeck, Xavier Lafosse, Jean-Luc Pelouard, and Riad Haïdar.   Total routing and absorption of photons in dual color plasmonic antennas. *Appl. Phys. Lett.*, 99(24):241104, Dec 2011. DOI: 10.1063/1.3670051 45

[102] Yong Ma, Qin Chen, James Grant, Shimul C. Saha, A. Khalid, and David R. S. Cumming. A terahertz polarization insensitive dual band metamaterial absorber. *Opt. Lett.*, 36(6):945, Mar 2011. DOI: 10.1364/ol.36.000945 45, 53

[103] J. Lee and S. Lim. Bandwidth-enhanced and polarisation-insensitive metamaterial absorber using double resonance. *Electron. Lett.*, 47(1):8, 2011. DOI: 10.1049/el.2010.2770 45

[104] Hua Cheng, Shuqi Chen, Haifang Yang, Junjie Li, Xin An, Changzhi Gu, and Jianguo Tian. A polarization insensitive and wide-angle dual-band nearly perfect absorber in the infrared regime. *J. Opt.*, 14(8):085102, Aug 2012. DOI: 10.1088/2040-8978/14/8/085102 45, 53, 54

[105] Li Huang, Dibakar Roy Chowdhury, Suchitra Ramani, Matthew T. Reiten, Sheng-Nian Luo, Antoinette J. Taylor, and Hou-Tong Chen. Experimental demonstration of terahertz metamaterial absorbers with a broad and flat high absorption band. *Opt. Lett.*, 37(2):154, Jan 2012. DOI: 10.1364/ol.37.000154 45

[106] Saptarshi Ghosh, Somak Bhattacharyya, Yadunath Kaiprath, and KuMar Vaibhav Srivastava. Bandwidth-enhanced polarization-insensitive microwave metamaterial absorber and its equivalent circuit model. *J. Appl. Phys.*, 115(10):104503, Mar 2014. DOI: 10.1063/1.4868577 45, 53, 54

[107] Riad Yahiaoui, Siyu Tan, Longqing Cong, RanJan Singh, Fengping Yan, and Weili Zhang. Multispectral terahertz sensing with highly flexible ultrathin metamaterial absorber. *J. Appl. Phys.*, 118(8):083103, Aug 2015. DOI: 10.1063/1.4929449 45

[108] Nanli Mou, Shulin Sun, Hongxing Dong, Shaohua Dong, Qiong He, Lei Zhou, and Long Zhang. Hybridization-induced broadband terahertz wave absorption with graphene metasurfaces. *Opt. Express*, 26(9):11728, Apr 2018. DOI: 10.1364/oe.26.011728 45

[109] Liang Huang and Hongsheng Chen. Multi-band and polarization insensitive Metamaterial absorber. *Prog. Electromagn. Res.*, 113:103–110, 2011. DOI: 10.2528/pier10122401 45

[110] Shuqi Chen, Hua Cheng, Haifang Yang, Junjie Li, Xiaoyang Duan, Changzhi Gu, and Jianguo Tian. Polarization insensitive and omnidirectional broadband near perfect planar metamaterial absorber in the near infrared regime. *Appl. Phys. Lett.*, 99(25):253104, Dec 2011. DOI: 10.1063/1.3670333 45

[111] Chihhui Wu and Gennady Shvets. Design of metamaterial surfaces with broadband absorbance. *Opt. Lett.*, 37(3):308, Feb 2012. DOI: 10.1364/ol.37.000308 45, 56

[112] Xiaopeng Shen, Yan Yang, Yuanzhang Zang, Jianqiang Gu, Jiaguang Han, Weili Zhang, and Tie Jun Cui. Triple-band terahertz metamaterial absorber: Design, experiment, and physical interpretation. *Appl. Phys. Lett.*, 101(15):154102, Oct 2012. DOI: 10.1063/1.4757879 45

[113] He-Xiu Xu, Guang-Ming Wang, Mei-Qing Qi, Jian-Gang Liang, Jian-Qiang Gong, and Zhi-Ming Xu. Triple-band polarization-insensitive wide-angle ultra-miniature metamaterial transmission line absorber. *Phys. Rev. B*, 86(20):205104, Nov 2012. DOI: 10.1103/physrevb.86.205104 45

[114] XiaoJun Huang, Helin Yang, Shengqing Yu, Jixin Wang, Minhua Li, and Qiwei Ye. Triple-band polarization-insensitive wide-angle ultra-thin planar spiral metamaterial absorber. *J. Appl. Phys.*, 113(21):213516, Jun 2013. DOI: 10.1063/1.4809655 45

[115] Huiqing Zhai, Chuanhan Zhan, Zhenhua Li, and Changhong Liang. A triple-band ultrathin metamaterial absorber With wide-angle and polarization stability. *IEEE Antennas Wirel. Propag. Lett.*, 14:241–244, 2015. DOI: 10.1109/lawp.2014.2361011 45

[116] Yanxia Cui, Jun Xu, Kin Hung Fung, Yi Jin, Anil KuMar, Sailing He, and Nicholas X. Fang. A thin film broadband absorber based on multi-sized nanoantennas. *Appl. Phys. Lett.*, 99(25):253101, Dec 2011. DOI: 10.1063/1.3672002 45

[117] B. X. Wang, X. Zhai, G. Z. Wang, W. Q. Huang, and L. L. Wang. Design of a four-band and polarization-insensitive terahertz metamaterial absorber. *IEEE Photonics J.*, 7(1):1–8, Feb 2015. DOI: 10.1109/jphot.2014.2381633 45

[118] Ben-Xin Wang. Quad-band terahertz metamaterial absorber based on the combining of the dipole and quadrupole resonances of two SRRs. *IEEE J. Sel. Top. Quantum Electron.*, 23(4):1–7, Jul 2017. DOI: 10.1109/jstqe.2016.2547325 45

[119] Jeremy A. Bossard, Lan Lin, Seokho Yun, Liu Liu, Douglas H. Werner, and Theresa S. Mayer. Near-ideal optical metamaterial absorbers with super-octave bandwidth. *ACS Nano*, 8(2):1517–1524, Feb 2014. DOI: 10.1021/nn4057148 45, 46, 47, 56, 57

[120] Mitchell Kenney, James Grant, Yash D. Shah, Ivonne Escorcia-Carranza, Mark Humphreys, and David R. S. Cumming. Octave-spanning broadband absorption of terahertz light using metasurface fractal-cross absorbers. *ACS Photonics*, 4(10):2604–2612, Oct 2017. DOI: 10.1021/acsphotonics.7b00906 45

[121] Hu Tao, Christopher M. Bingham, Daniel V. Pilon, Kebin Fan, Andrew C. Strikwerda, David Shrekenhamer, Willie J. Padilla, Xin Zhang, and Richard D. Averitt. A dual band terahertz metamaterial absorber. *J. Phys. D. Appl. Phys.*, 43(22):225102, Jun 2010. DOI: 10.1088/0022-3727/43/22/225102 46

[122] Minhua Li, He-Lin Yang, Xi-Wen Hou, Yan Tian, and Dong-Yun Hou. Perfect metamaterial absorber with dual bands. *Prog. Electromagn. Res.*, 108:37–49, 2010. DOI: 10.2528/pier10071409 45

[123] Somak Bhattacharyya, Saptarshi Ghosh, and KuMar Vaibhav Srivastava. Triple band polarization-independent metamaterial absorber with bandwidth enhancement at X-band. *J. Appl. Phys.*, 114(9):094514, Sep 2013. DOI: 10.1063/1.4820569 45

[124] Somak Bhattacharyya and KuMar Vaibhav Srivastava. Triple band polarization-independent ultra-thin metamaterial absorber using electric field-driven LC resonator. *J. Appl. Phys.*, 115(6):064508, Feb 2014. DOI: 10.1063/1.4865273 45

[125] Bingxin Zhang, Yanhui Zhao, Qingzhen Hao, Brian Kiraly, Iam-Choon Khoo, Shufen Chen, and Tony Jun Huang. Polarization-independent dual-band infrared perfect absorber based on a metal-dielectric-metal elliptical nanodisk array. *Opt. Express*, 19(16):15221, Aug 2011. DOI: 10.1364/oe.19.015221 45, 56

[126] Zhi Hao Jiang, Seokho Yun, Fatima Toor, Douglas H. Werner, and Theresa S. Mayer. Conformal dual-band near-perfectly absorbing mid-infrared metamaterial coating. *ACS Nano*, 5(6):4641–4647, Jun 2011. DOI: 10.1021/nn2004603 45, 60

[127] Kamil Boratay Alici, Adil Burak Turhan, Costas M. Soukoulis, and Ekmel Ozbay. Optically thin composite resonant absorber at the near-infrared band: a polarization independent and spectrally broadband configuration. *Opt. Express*, 19(15):14260, Jul 2011. DOI: 10.1364/oe.19.014260 45

[128] Abul K. Azad, Wilton J. M. Kort-Kamp, Milan Sykora, Nina R. Weisse-Bernstein, Ting S. Luk, Antoinette J. Taylor, Diego A. R. Dalvit, and Hou-Tong Chen. Metasurface broadband solar absorber. *Sci. Rep.*, 6(1):20347, Apr 2016. DOI: 10.1038/srep20347 47

[129] Patrick Bouchon, Charlie Koechlin, Fabrice Pardo, Riad Haïdar, and Jean-Luc Pelouard. Wideband omnidirectional infrared absorber with a patchwork of plasmonic nanoantennas. *Opt. Lett.*, 37(6):1038, Mar 2012. DOI: 10.1364/ol.37.001038 47

[130] Michael G. Nielsen, Anders Pors, Ole Albrektsen, and Sergey I. Bozhevolnyi. Efficient absorption of visible radiation by gap plasmon resonators. *Opt. Express*, 20(12):13311, Jun 2012. DOI: 10.1364/oe.20.013311 47

[131] Cheng-Wen Cheng, Mohammed Nadhim Abbas, Chao-Wei Chiu, Kun-Ting Lai, Min-Hsiung Shih, and Yia-Chung Chang. Wide-angle polarization independent infrared broadband absorbers based on metallic multi-sized disk arrays. *Opt. Express*, 20(9):10376, Apr 2012. DOI: 10.1364/oe.20.010376 47, 53

[132] Yahong Liu, Shuai Gu, Chunrong Luo, and Xiaopeng Zhao. Ultra-thin broadband metamaterial absorber. *Appl. Phys. A*, 108(1):19–24, Jul 2012. DOI: 10.1007/s00339-012-6936-0 47

[133] Nan Zhang, Peiheng Zhou, Dengmu Cheng, Xiaolong Weng, Jianliang Xie, and Longjiang Deng.   Dual-band absorption of mid-infrared metamaterial absorber based on distinct dielectric spacing layers. *Opt. Lett.*, 38(7):1125, Apr 2013. DOI: 10.1364/ol.38.001125 47, 56

[134] Jin Woo Park, Pham Van Tuong, Joo Yull Rhee, Ki Won Kim, Won Ho Jang, Eun Ha Choi, Liang Yao Chen, and YoungPak Lee. Multi-band metamaterial absorber based on the arrangement of donut-type resonators. *Opt. Express*, 21(8):9691, Apr 2013. DOI: 10.1364/oe.21.009691 47

[135] Boyang Zhang, Joshua Hendrickson, and Junpeng Guo.   Multispectral near-perfect metamaterial absorbers using spatially multiplexed plasmon resonance metal square structures. *J. Opt. Soc. Am. B*, 30(3):656, Mar 2013. DOI: 10.1364/josab.30.000656 47, 56

[136] Hao Wang and Liping Wang. Perfect selective metamaterial solar absorbers. *Opt. Express*, 21(S6):A1078, Nov 2013. DOI: 10.1364/oe.21.0a1078 47

[137] Wei Ma, Yongzheng Wen, and Xiaomei Yu. Broadband metamaterial absorber at mid-infrared using multiplexed cross resonators. *Opt. Express*, 21(25):30724, Dec 2013. DOI: 10.1364/oe.21.030724 47, 56

[138] Ximin Tian and Zhi-Yuan Li.   Visible-near infrared ultra-broadband polarization-independent metamaterial perfect absorber involving phase-change materials. *Photonics Res.*, 4(4):146, Aug 2016. DOI: 10.1364/prj.4.000146 47, 56

[139] Xiaoming Liu, Chuwen Lan, Bo Li, Qian Zhao, and Ji Zhou. Dual band metamaterial perfect absorber based on artificial dielectric "molecules". *Sci. Rep.*, 6(1):28906, Sep 2016. DOI: 10.1038/srep28906 47

[140] Zhigang Li, Liliana Stan, David A. Czaplewski, Xiaodong Yang, and Jie Gao. Wavelength-selective mid-infrared metamaterial absorbers with multiple tungsten cross resonators. *Opt. Express*, 26(5):5616, Mar 2018. DOI: 10.1364/oe.26.005616 47, 56, 59

[141] Joshua Hendrickson, Junpeng Guo, Boyang Zhang, Walter Buchwald, and Richard Soref.   Wideband perfect light absorber at midwave infrared using multiplexed metal structures. *Opt. Lett.*, 37(3):371, Feb 2012. DOI: 10.1364/ol.37.000371 47

[142] Zhigang Li, Liliana Stan, David A. Czaplewski, Xiaodong Yang, and Jie Gao. Broadband infrared binary-pattern metasurface absorbers with micro-genetic algorithm optimization. *Opt. Lett.*, 44(1):114, Jan 2019. DOI: 10.1364/ol.44.000114 47, 59

[143] J.S. Jensen and O. Sigmund. Topology optimization for nano-photonics. *Laser & Photonics Reviews*, 5(2):308–321, December 2010. DOI: 10.1002/lpor.201000014 47

[144] Christopher M. Lalau-Keraly, Samarth Bhargava, Owen D. Miller, and Eli Yablonovitch. Adjoint shape optimization applied to electromagnetic design. *Optics Express*, 21(18):21693, September 2013. DOI: 10.1364/oe.21.021693 47

[145] Sze Ming Fu, Yan Kai Zhong, Nyan Ping Ju, Ming-Hsiang Tu, Bo-Ruei Chen, and Albert Lin. Broadband polarization-insensitive metamaterial perfect absorbers using topology optimization. *IEEE Photonics Journal*, 8(5):1–11, October 2016. DOI: 10.1109/jphot.2016.2602335 47

[146] Zhaocheng Liu, Dayu Zhu, Sean P. Rodrigues, Kyu-Tae Lee, and Wenshan Cai. Generative model for the inverse design of metasurfaces. *Nano Lett.*, 18(10):6570–6576, Oct 2018. DOI: 10.1021/acs.nanolett.8b03171 47

[147] Dianjing Liu, Yixuan Tan, Erfan Khoram, and Zongfu Yu. Training deep neural networks for the inverse design of nanophotonic structures. *ACS Photonics*, 5(4):1365–1369, Apr 2018. DOI: 10.1021/acsphotonics.7b01377 47

[148] Itzik Malkiel, Michael Mrejen, Achiya Nagler, Uri Arieli, Lior Wolf, and Haim Suchowski. Plasmonic nanostructure design and characterization via Deep Learning. *Light Sci. Appl.*, 7(1):60, Dec 2018. DOI: 10.1038/s41377-018-0060-7 47

[149] Wei Ma, Feng Cheng, and Yongmin Liu. Deep-learning-enabled on-demand design of chiral metamaterials. *ACS Nano*, 12(6):6326–6334, Jun 2018. DOI: 10.1021/acsnano.8b03569 47

[150] Qian Zhang, Che Liu, Xiang Wan, Lei Zhang, Shuo Liu, Yan Yang, and Tie Jun Cui. Machine-learning designs of anisotropic digital coding metasurfaces. *Adv. Theory Simulations*, 2(2):1800132, Feb 2019. DOI: 10.1002/adts.201800132 47

[151] Christopher M. Bingham, Hu Tao, Xianliang Liu, Richard D. Averitt, Xin Zhang, and Willie J. Padilla. Planar wallpaper group metamaterials for novel terahertz applications. *Opt. Express*, 16(23):18565–18575, Nov 2008. DOI: 10.1364/oe.16.018565 47

[152] Yu Qian Ye, Yi Jin, and Sailing He. Omnidirectional, polarization-insensitive and broadband thin absorber in the terahertz regime. *J. Opt. Soc. Am. B*, 27(3):498, Mar 2010. DOI: 10.1364/josab.27.000498 48, 49

[153] James Grant, Yong Ma, Shimul Saha, Ata Khalid, and David R. S. Cumming. Polarization insensitive, broadband terahertz metamaterial absorber. *Opt. Lett.*, 36(17):3476, Sep 2011. DOI: 10.1364/ol.36.003476 48, 52

[154] Fangrong Hu, Li Wang, Baogang Quan, Xinlong Xu, Zhi Li, Zhongan Wu, and Xuecong Pan. Design of a polarization insensitive multiband terahertz metamaterial absorber. *J. Phys. D. Appl. Phys.*, 46(19):195103, May 2013. DOI: 10.1088/0022-3727/46/19/195103 48

[155] Muhammad Amin, Mohamed Farhat, and Hakan Bağcı. An ultra-broadband multilayered graphene absorber. *Opt. Express*, 21(24):29938, Dec 2013. DOI: 10.1364/oe.21.029938 48, 50

[156] Han Xiong, Jin-Song Hong, Chao-Ming Luo, and Lin-Lin Zhong. An ultrathin and broadband metamaterial absorber using multi-layer structures. *J. Appl. Phys.*, 114(6):064109, Aug 2013. DOI: 10.1063/1.4818318 48

[157] Ben-Xin Wang, Ling-Ling Wang, Gui-Zhen Wang, Wei-Qing Huang, Xiao-Fei Li, and Xiang Zhai. Theoretical investigation of broadband and wide-angle terahertz metamaterial absorber. *IEEE Photonics Technol. Lett.*, 26(2):111–114, Jan 2014. DOI: 10.1109/lpt.2013.2289299 48

[158] Jianfei Zhu, Zhaofeng Ma, Wujiong Sun, Fei Ding, Qiong He, Lei Zhou, and Yungui Ma. Ultra-broadband terahertz metamaterial absorber. *Appl. Phys. Lett.*, 105(2):021102, Jul 2014. DOI: 10.1063/1.4890521 48

[159] Zhaoxian Su, Jianbo Yin, and Xiaopeng Zhao. Terahertz dual-band metamaterial absorber based on graphene/MgF_2 multilayer structures. *Opt. Express*, 23(2):1679, Jan 2015. DOI: 10.1364/oe.23.001679 48

[160] Shuo Liu, Haibing Chen, and Tie Jun Cui. A broadband terahertz absorber using multi-layer stacked bars. *Appl. Phys. Lett.*, 106(15):151601, Apr 2015. DOI: 10.1063/1.4918289 48

[161] Sailing He, Fei Ding, Lei Mo, and Fanglin Bao. Light absorber with an ultra-broad flat band based on multi-sized slow-wave hyperbolic metamaterial thin-films (invited paper). *Prog. Electromagn. Res.*, 147:69–79, 2014. DOI: 10.2528/pier14040306 48, 49

[162] Han Lin, Björn C. P. Sturmberg, Keng-Te Lin, Yunyi Yang, Xiaorui Zheng, Teck K. Chong, C. Martijn de Sterke, and Baohua Jia. A 90-nm-thick graphene metamaterial for strong and extremely broadband absorption of unpolarized light. *Nat. Photonics*, 13(4):270–276, Apr 2019. DOI: 10.1038/s41566-019-0389-3 48, 49, 50

[163] Fei Ding, Yanxia Cui, Xiaochen Ge, Yi Jin, and Sailing He. Ultra-broadband microwave metamaterial absorber. *Appl. Phys. Lett.*, 100(10):103506, Mar 2012. DOI: 10.1063/1.3692178 48

[164] Wenrui Xue, Xi Chen, Yanling Peng, and Rongcao Yang. Grating-type mid-infrared light absorber based on silicon carbide material. *Opt. Express*, 24(20):22596, Oct 2016. DOI: 10.1364/oe.24.022596 48

[165] Jiao Wang and Yannan Jiang. Infrared absorber based on sandwiched two-dimensional black phosphorus metamaterials. *Opt. Express*, 25(5):5206, Mar 2017. DOI: 10.1364/oe.25.005206 48

[166] Caner Guclu, Salvatore Campione, and Filippo Capolino. Hyperbolic metamaterial as super absorber for scattered fields generated at its surface. *Phys. Rev. B*, 86(20):205130, Nov 2012. DOI: 10.1103/physrevb.86.205130 48

[167] Sailing He and Tuo Chen. Broadband THz absorbers with graphene-based anisotropic metamaterial films. *IEEE Trans. Terahertz Sci. Technol.*, 3(6):757–763, Nov 2013. DOI: 10.1109/tthz.2013.2283370 48, 50

[168] Dengxin Ji, Haomin Song, Xie Zeng, Haifeng Hu, Kai Liu, Nan Zhang, and Qiao-qiang Gan. Broadband absorption engineering of hyperbolic metafilm patterns. *Sci. Rep.*, 4(1):4498, May 2014. DOI: 10.1038/srep04498 48

[169] Jing Zhou, Alexander F. Kaplan, Long Chen, and L. Jay Guo. Experiment and theory of the broadband absorption by a tapered hyperbolic metamaterial array. *ACS Photonics*, 1(7):618–624, Jul 2014. DOI: 10.1021/ph5001007 48

[170] Fei Ding, Yi Jin, Borui Li, Hao Cheng, Lei Mo, and Sailing He. Ultrabroadband strong light absorption based on thin multilayered metamaterials. *Laser Photon. Rev.*, 8(6):946–953, Nov 2014. DOI: 10.1002/lpor.201400157 48

[171] F. Bonaccorso, Z. Sun, T. Hasan, and A. C. Ferrari. Graphene photonics and opto-electronics. *Nat. Photonics*, 4(9):611–622, Sep 2010. DOI: 10.1038/nphoton.2010.186 50

[172] Qing Hua Wang, Kourosh Kalantar-Zadeh, Andras Kis, Jonathan N. Coleman, and Michael S. Strano. Electronics and optoelectronics of two-dimensional tran-sition metal dichalcogenides. *Nat. Nanotechnol.*, 7(11):699–712, Nov 2012. DOI: 10.1038/nnano.2012.193 50

[173] Likai Li, YiJun Yu, Guo Jun Ye, Qingqin Ge, Xuedong Ou, Hua Wu, Donglai Feng, Xian Hui Chen, and Yuanbo Zhang. Black phosphorus field-effect transistors. *Nat. Nanotechnol.*, 9(5):372–377, May 2014. DOI: 10.1038/nnano.2014.35 50

[174] Mohamed A. K. Othman, Caner Guclu, and Filippo Capolino. Graphene-based tunable hyperbolic metamaterials and enhanced near-field absorption. *Opt. Express*, 21(6):7614, Mar 2013. DOI: 10.1364/oe.21.007614 50

[175] Renxia Ning, Shaobin Liu, Haifeng Zhang, and Zheng Jiao. Dual-gated tunable ab-sorption in graphene-based hyperbolic metamaterial. *AIP Adv.*, 5(6):067106, Jun 2015. DOI: 10.1063/1.4922170 50

[176] S. Gu, J. P. Barrett, T. H. Hand, B.-I. Popa, and S. A. Cummer. A broadband low-reflection metamaterial absorber. *J. Appl. Phys.*, 108(6):064913, Sep 2010. DOI: 10.1063/1.3485808 50

[177] Yong Zhi Cheng, Ying Wang, Yan Nie, Rong Zhou Gong, Xuan Xiong, and Xian Wang. Design, fabrication and measurement of a broadband polarization-insensitive metamaterial absorber based on lumped elements. *J. Appl. Phys.*, 111(4):044902, Feb 2012. DOI: 10.1063/1.3684553 50, 53, 54

[178] Sijia Li, Jun Gao, Xiangyu Cao, Wenqiang Li, Zhao Zhang, and Di Zhang. Wideband, thin, and polarization-insensitive perfect absorber based the double octagonal rings metamaterials and lumped resistances. *J. Appl. Phys.*, 116(4):043710, Jul 2014. DOI: 10.1063/1.4891716 50, 53

[179] Kai Chen, Thang Duy Dao, Satoshi Ishii, Masakazu Aono, and Tadaaki Nagao. Infrared aluminum metamaterial perfect absorbers for plasmon-enhanced infrared spectroscopy. *Adv. Funct. Mater.*, 25(42):6637–6643, Nov 2015. DOI: 10.1002/adfm.201501151 50, 56

[180] Xingxing Chen, Hanmo Gong, Shuowei Dai, Ding Zhao, Yuanqing Yang, Qiang Li, and Min Qiu. Near-infrared broadband absorber with film-coupled multilayer nanorods. *Opt. Lett.*, 38(13):2247, Jul 2013. DOI: 10.1364/ol.38.002247 51

[181] R.C. McPhedran, L.C. Botten, M.S. Craig, M. Nevière, and D. Maystre. Lossy lamellar gratings in the quasistatic limit. *Optica Acta: International Journal of Optics*, 29(3):289–312, March 1982. DOI: 10.1080/713820844 51

[182] L. C. Botten, R. C. McPhedran, N. A. Nicorovici, and G. H. Derrick. Periodic models for thin optimal absorbers of electromagnetic radiation. *Physical Review B*, 55(24):R16072–R16082, June 1997. DOI: 10.1103/physrevb.55.r16072 51

[183] E. Popov, D. Maystre, R. C. McPhedran, M. Nevière, M.C. Hutley, and G. H. Derrick. Total absorption of unpolarized light by crossed gratings. *Opt. Express*, 16(9):6146, Apr 2008. DOI: 10.1364/oe.16.006146 51

[184] Jean-Jacques Greffet, Rémi Carminati, Karl Joulain, Jean-Philippe Mulet, Stéphane Mainguy, and Yong Chen. Coherent emission of light by thermal sources. *Nature*, 416(6876):61–64, Mar 2002. DOI: 10.1038/416061a 51

[185] Lijun Meng, Ding Zhao, Zhichao Ruan, Qiang Li, Yuanqing Yang, and Min Qiu. Optimized grating as an ultra-narrow band absorber or plasmonic sensor. *Opt. Lett.*, 39(5):1137, Mar 2014. DOI: 10.1364/ol.39.001137 51

[186] Sungho Kang, Zhenyun Qian, Vageeswar Rajaram, Sila Deniz Calisgan, Andrea Alù, and Matteo Rinaldi. Ultra-narrowband metamaterial absorbers for high spectral resolution infrared spectroscopy. *Adv. Opt. Mater.*, 7(2):1801236, Jan 2019. DOI: 10.1002/adom.201801236 51, 52

[187] Jinna He, Pei Ding, Junqiao Wang, Chunzhen Fan, and Erjun Liang. Ultra-narrow band perfect absorbers based on plasmonic analog of electromagnetically induced absorption. *Opt. Express*, 23(5):6083, Mar 2015. DOI: 10.1364/oe.23.006083 51

[188] Ricardo Marqués, Francisco Medina, and Rachid Rafii-El-Idrissi. Role of bianisotropy in negative permeability and left-handed metamaterials. *Phys. Rev. B*, 65(14):144440, Apr 2002. DOI: 10.1103/physrevb.65.144440 52

[189] T J Yen, W J Padilla, N Fang, D C Vier, D R Smith, J B Pendry, D N Basov, and X Zhang. Terahertz magnetic response from artificial materials. *Science*, 303(5663):1494–6, Mar 2004. DOI: 10.1126/science.1094025 52

[190] W. J. Padilla, A. J. Taylor, and R. D. Averitt. Dynamical electric and magnetic metamaterial response at terahertz frequencies. *Phys. Rev. Lett.*, 96(10):107401, Mar 2006. DOI: 10.1103/physrevlett.96.107401 52

[191] W. Padilla, M. Aronsson, C. Highstrete, Mark Lee, A. Taylor, and R. Averitt. Electrically resonant terahertz metamaterials: Theoretical and experimental investigations. *Phys. Rev. B*, 75(4):041102, Jan 2007. DOI: 10.1103/physrevb.75.041102 52

[192] Willie J. Padilla. Group theoretical description of artificial electromagnetic metamaterials. *Opt. Express*, 15(4):1639, Feb 2007. DOI: 10.1364/oe.15.001639 52

[193] Zhu Bo, Wang Zheng-Bin, Yu Zhen-Zhong, Zhang Qi, Zhao Jun-Ming, Feng Yi-Jun, and Jiang Tian. Planar metamaterial microwave absorber for all wave polarizations. *Chinese Phys. Lett.*, 26(11):114102, Nov 2009. DOI: 10.1088/0256-307x/26/11/114102 52

[194] Chao Gu, Shao-Bo Qu, Zhi-Bin Pei, and Zhuo Xu. A metamaterial absorber with direction-selective and polarisation-insensitive properties. *Chinese Phys. B*, 20(3):037801, Mar 2011. DOI: 10.1088/1674-1056/20/3/037801 52

[195] P. V. Tuong, J. W. Park, J. Y. Rhee, K. W. Kim, W. H. Jang, H. Cheong, and Y. P. Lee. Polarization-insensitive and polarization-controlled dual-band absorption in metamaterials. *Appl. Phys. Lett.*, 102(8):081122, Feb 2013. DOI: 10.1063/1.4794173 52

[196] Bingnan Wang, Thomas Koschny, and Costas M. Soukoulis. Wide-angle and polarization-independent chiral metamaterial absorber. *Phys. Rev. B*, 80(3):033108, Jul 2009. DOI: 10.1103/physrevb.80.033108 52, 53, 54

[197] Long Li, Yang Yang, and Changhong Liang. A wide-angle polarization-insensitive ultra-thin metamaterial absorber with three resonant modes. *J. Appl. Phys.*, 110(6):063702, Sep 2011. DOI: 10.1063/1.3638118 53, 54, 56

[198] Weiren Zhu, Xiaopeng Zhao, Boyi Gong, Longhai Liu, and Bin Su. Optical metamaterial absorber based on leaf-shaped cells. *Appl. Phys. A*, 102(1):147–151, Jan 2011. DOI: 10.1007/s00339-010-6057-6 53

[199] Dongju Lee, Jung Gyu Hwang, Daecheon Lim, Tadayoshi Hara, and Sungjoon Lim. Incident angle- and polarization-insensitive metamaterial absorber using circular sectors. *Sci. Rep.*, 6(1):27155, Jul 2016. DOI: 10.1038/srep27155 53

[200] Xun-Jun He, Yue Wang, Jianmin Wang, Tailong Gui, and Qun Wu. Dual-band terahertz metamaterial absorber with polarization insensitivity and wide incident angle. *Prog. Electromagn. Res.*, 115:381–397, 2011. DOI: 10.2528/pier11022307 53

[201] Meng Wu, Xiaoguang Zhao, Jingdi Zhang, Jacob Schalch, Guangwu Duan, Kevin Cremin, Richard D. Averitt, and Xin Zhang. A three-dimensional all-metal terahertz metamaterial perfect absorber. *Appl. Phys. Lett.*, 111(5):051101, Jul 2017. DOI: 10.1063/1.4996897 53

[202] Chihhui Wu, Burton Neuner, Gennady Shvets, Jeremy John, Andrew Milder, Byron Zollars, and Steve Savoy. Large-area wide-angle spectrally selective plasmonic absorber. *Phys. Rev. B*, 84(7):075102, Aug 2011. DOI: 10.1103/physrevb.84.075102 53

[203] Mingbo Pu, Chenggang Hu, Min Wang, Cheng Huang, Zeyu Zhao, Changtao Wang, Qin Feng, and Xiangang Luo. Design principles for infrared wide-angle perfect absorber based on plasmonic structure. *Opt. Express*, 19(18):17413, Aug 2011. DOI: 10.1364/oe.19.017413 53

[204] Pin Chieh Wu, Chun Yen Liao, Jia-Wern Chen, and Din Ping Tsai. Isotropic absorption and sensor of vertical split-ring resonator. *Adv. Opt. Mater.*, 5(2):1600581, Jan 2017. DOI: 10.1002/adom.201600581 53

[205] Zheyu Fang, Yu-Rong Zhen, Linran Fan, Xing Zhu, and Peter Nordlander. Tunable wide-angle plasmonic perfect absorber at visible frequencies. *Phys. Rev. B*, 85(24):245401, Jun 2012. DOI: 10.1103/physrevb.85.245401 53

[206] Wen-Chen Chen, Andrew Cardin, Machhindra Koirala, Xianliang Liu, Talmage Tyler, Kevin G. West, Christopher M. Bingham, Tatiana Starr, Anthony F. Starr, Nan M. Jokerst, and Willie J. Padilla. Role of surface electromagnetic waves in metamaterial absorbers. *Opt. Express*, 24(6):6783, Mar 2016. DOI: 10.1364/oe.24.006783 53

[207] Yu Tong Zhao, Bian Wu, Bei Ju Huang, and Qiang Cheng. Switchable broadband terahertz absorber/reflector enabled by hybrid graphene-gold metasurface. *Opt. Express*, 25(7):7161, Apr 2017. DOI: 10.1364/oe.25.007161 54

[208] Michelle C. Sherrott, Philip W. C. Hon, Katherine T. Fountaine, Juan C. Garcia, Samuel M. Ponti, Victor W. Brar, Luke A. Sweatlock, and Harry A. Atwater. Experimental demonstration of >230° phase modulation in gate-tunable graphene–gold reconfigurable mid-infrared metasurfaces. *Nano Lett.*, 17(5):3027–3034, May 2017. DOI: 10.1021/acs.nanolett.7b00359 54, 56

[209] Shinpei Ogawa, Daisuke Fujisawa, Hisatoshi Hata, Mitsuharu Uetsuki, Koji Misaki, and Masafumi Kimata. Mushroom plasmonic metamaterial infrared absorbers. *Appl. Phys. Lett.*, 106(4):041105, Jan 2015. DOI: 10.1063/1.4906860 54

[210] Jing Wang, Yiting Chen, Jiaming Hao, Min Yan, and Min Qiu. Shape-dependent absorption characteristics of three-layered metamaterial absorbers at near-infrared. *J. Appl. Phys.*, 109(7):074510, Apr 2011. DOI: 10.1063/1.3573495 55, 59

[211] Takahiro Yokoyama, Thang Duy Dao, Kai Chen, Satoshi Ishii, Ramu Pasupathi Sugavaneshwar, Masahiro Kitajima, and Tadaaki Nagao. Spectrally selective mid-infrared thermal emission from molybdenum plasmonic metamaterial operated up to 1000 °C. *Adv. Opt. Mater.*, 4(12):1987–1992, Dec 2016. DOI: 10.1002/adom.201600455 55, 56, 59

[212] Kai Chen, Ronen Adato, and Hatice Altug. Dual-band perfect absorber for multispectral plasmon-enhanced infrared spectroscopy. *ACS Nano*, 6(9):7998–8006, Sep 2012. DOI: 10.1021/nn3026468 55

[213] Matthew J. Rozin, David A. Rosen, Tyler J. Dill, and Andrea R. Tao. Colloidal metasurfaces displaying near-ideal and tunable light absorbance in the infrared. *Nat. Commun.*, 6(1):7325, Dec 2015. DOI: 10.1038/ncomms8325 55, 56

[214] Junghyun Park, Ju-Hyung Kang, Soo Jin Kim, Xiaoge Liu, and Mark L. Brongersma. Dynamic reflection phase and polarization control in metasurfaces. *Nano Lett.*, 17(1):407–413, Jan 2017. DOI: 10.1021/acs.nanolett.6b04378 55

[215] Jon W. Stewart, Gleb M. Akselrod, David R. Smith, and Maiken H. Mikkelsen. Toward multispectral imaging with colloidal metasurface pixels. *Adv. Mater.*, 29(6):1602971, Feb 2017. DOI: 10.1002/adma.201602971 55

[216] Zuojia Wang, Hui Jia, Kan Yao, Wenshan Cai, Hongsheng Chen, and Yongmin Liu. Circular dichroism metamirrors with near-perfect extinction. *ACS Photonics*, 3(11):2096–2101, Nov 2016. DOI: 10.1021/acsphotonics.6b00533 53

[217] Guy Lipworth, Nicholas W. Caira, Stéphane Larouche, and David R. Smith. Phase and magnitude constrained metasurface holography at W-band frequencies. *Opt. Express*, 24(17):19372, Aug 2016. DOI: 10.1364/oe.24.019372 53

[218] Xiaoguang Zhao, Kebin Fan, Jingdi Zhang, Huseyin R. Seren, Grace D. Metcalfe, Michael Wraback, Richard D. Averitt, and Xin Zhang. Optically tunable metamaterial perfect absorber on highly flexible substrate. *Sensors Actuators A Phys.*, 231:74–80, Jul 2015. DOI: 10.1016/j.sna.2015.02.040 53

[219] Xi Chen, Yiting Chen, Min Yan, and Min Qiu. Nanosecond photothermal effects in plasmonic nanostructures. *ACS Nano*, 6(3):2550–2557, Mar 2012. DOI: 10.1021/nn2050032 56

[220] Borislav Vasić and Radoš Gajić. Graphene induced spectral tuning of metamaterial absorbers at mid-infrared frequencies. *Appl. Phys. Lett.*, 103(26):261111, Dec 2013. DOI: 10.1063/1.4858459 56

[221] Yuping Zhang, Tongtong Li, Qi Chen, Huiyun Zhang, John F. O'Hara, Ethan Abele, Antoinette J. Taylor, Hou-Tong Chen, and Abul K. Azad. Independently tunable dual-band perfect absorber based on graphene at mid-infrared frequencies. *Sci. Rep.*, 5(1):18463, Nov 2016. DOI: 10.1038/srep18463 56

[222] Joshua A. Mason, Graham Allen, Viktor A. Podolskiy, and Daniel Wasserman. Strong coupling of molecular and mid-infrared perfect absorber resonances. *IEEE Photonics Technol. Lett.*, 24(1):31–33, Jan 2012. DOI: 10.1109/lpt.2011.2171942 56

[223] James Grant, Iain J. H. McCrindle, and David R. S. Cumming. Multi-spectral materials: hybridisation of optical plasmonic filters, a mid infrared metamaterial absorber and a terahertz metamaterial absorber. *Opt. Express*, 24(4):3451, Feb 2016. DOI: 10.1364/oe.24.003451 56

[224] Yurui Qu, Qiang Li, Kaikai Du, Lu Cai, Jun Lu, and Min Qiu. Dynamic thermal emission control based on ultrathin plasmonic metamaterials including phase-changing material GST. *Laser Photon. Rev.*, 11(5):1700091, Sep 2017. DOI: 10.1002/lpor.201700091 56

[225] Cheng Shi, Nathan H. Mahlmeister, Isaac J. Luxmoore, and Geoffrey R. Nash. Metamaterial-based graphene thermal emitter. *Nano Res.*, 11(7):3567–3573, Jul 2018. DOI: 10.1007/s12274-017-1922-7 56

[226] Beibei Zeng, Zhiqin Huang, Akhilesh Singh, Yu Yao, Abul K. Azad, Aditya D. Mohite, Antoinette J. Taylor, David R. Smith, and Hou-Tong Chen. Hybrid graphene metasurfaces for high-speed mid-infrared light modulation and single-pixel imaging. *Light Sci. Appl.*, 7(1):51, Dec 2018. DOI: 10.1038/s41377-018-0055-4 56

[227] Alireza Bonakdar, Mohsen Rezaei, Eric Dexheimer, and Hooman Mohseni. High-throughput realization of an infrared selective absorber/emitter by DUV microsphere

projection lithography. *Nanotechnology*, 27(3):035301, Jan 2016. DOI: 10.1088/0957-4484/27/3/035301 56

[228] Mohammadamir Ghaderi, Ehsan Karimi ShahMarvandi, and Reinoud F. Wolffenbuttel. CMOS-compatible mid-IR metamaterial absorbers for out-of-band suppression in optical MEMS. *Optical Materials Express*, 8(7):1696, Jul 2018. DOI: 10.1364/ome.8.001696 56

[229] Prakash Pitchappa, Chong Pei Ho, Piotr Kropelnicki, Navab Singh, Dim-Lee Kwong, and Chengkuo Lee. Micro-electro-mechanically switchable near infrared complementary metamaterial absorber. *Appl. Phys. Lett.*, 104(20):201114, May 2014. DOI: 10.1063/1.4879284 56

[230] Gleb M. Akselrod, Jiani Huang, Thang B. Hoang, Patrick T. Bowen, Logan Su, David R. Smith, and Maiken H. Mikkelsen. Large-area metasurface perfect absorbers from visible to near-infrared. *Adv. Mater.*, 27(48):8028–8034, Dec 2015. DOI: 10.1002/adma.201503281 56, 60

[231] Mehdi Keshavarz Hedayati, Mojtaba Javaherirahim, Babak Mozooni, Ramzy Abdelaziz, Ali Tavassolizadeh, Venkata Sai Kiran Chakravadhanula, Vladimir Zaporojtchenko, Thomas Strunkus, Franz Faupel, and Mady Elbahri. Design of a perfect black absorber at visible frequencies using plasmonic metamaterials. *Adv. Mater.*, 23(45):5410–5414, Dec 2011. DOI: 10.1002/adma.201102646 56

[232] Zhengqi Liu, Xiaoshan Liu, Shan Huang, Pingping Pan, Jing Chen, Guiqiang Liu, and Gang Gu. Automatically acquired broadband plasmonic-metamaterial black absorber during the metallic film-formation. *ACS Appl. Mater. Interfaces*, 7(8):4962–4968, Mar 2015. DOI: 10.1021/acsami.5b00056 56

[233] Hui Li, Li Hua Yuan, Bin Zhou, Xiao Peng Shen, Qiang Cheng, and Tie Jun Cui. Ultrathin multiband gigahertz metamaterial absorbers. *J. Appl. Phys.*, 110(1):014909, Jul 2011. DOI: 10.1063/1.3608246 56

[234] Bo Zhu, Zhengbin Wang, Ci Huang, Yijun Feng, Junming Zhao, and Tian Jiang. Polarization insensitive metamaterial absorber with wide incident angle. *Prog. Electromagn. Res.*, 101:231–239, 2010. DOI: 10.2528/pier10011110 56

[235] W.-C. Chen, C. M. Bingham, K. M. Mak, N. W. Caira, and W. J. Padilla. Extremely subwavelength planar magnetic metamaterials. *Phys. Rev. B*, 85:201104, May 2012. DOI: 10.1103/physrevb.85.201104 56

[236] Kamil Boratay Alici, Filiberto Bilotti, Lucio Vegni, and Ekmel Ozbay. Experimental verification of metamaterial based subwavelength microwave absorbers. *J. Appl. Phys.*, 108(8):083113, Oct 2010. DOI: 10.1063/1.3493736 56

[237] Yongzhi Cheng, Helin Yang, Zhengze Cheng, and Nan Wu. Perfect metamaterial absorber based on a split-ring-cross resonator. *Appl. Phys. A*, 102(1):99–103, Jan 2011. DOI: 10.1007/s00339-010-6022-4 56

[238] Christian C. Nadell, Bohao Huang, Jordan M. Malof, and Willie J. Padilla. Deep learning for accelerated all-dielectric metasurface design. *Optics Express*, 27(20):27523, September 2019. DOI: 10.1364/oe.27.027523 56

[239] Yang Wang, Tianyi Sun, Trilochan Paudel, Yi Zhang, Zhifeng Ren, and Krzysztof Kempa. Metamaterial-plasmonic absorber structure for high efficiency amorphous silicon solar cells. *Nano Lett.*, 12(1):440–445, Jan 2012. DOI: 10.1021/nl203763k 57, 58

[240] Dongxing Wang, Wenqi Zhu, Michael D. Best, Jon P. Camden, and Kenneth B. Crozier. Wafer-scale metasurface for total power absorption, local field enhancement and single molecule Raman spectroscopy. *Sci. Rep.*, 3(1):2867, Dec 2013. DOI: 10.1038/srep02867 57, 58

[241] Riad Yahiaoui, Jean Paul Guillet, Frédérick de Miollis, and Patrick Mounaix. Ultra-flexible multiband terahertz metamaterial absorber for conformal geometry applications. *Opt. Lett.*, 38(23):4988, Dec 2013. DOI: 10.1364/ol.38.004988 57

[242] Iain J. H. McCrindle, James Grant, Timothy D. Drysdale, and David R. S. Cumming. Multi-spectral materials: hybridisation of optical plasmonic filters and a terahertz metamaterial absorber. *Adv. Opt. Mater.*, 2(2):149–153, Feb 2014. DOI: 10.1002/adom.201300408 57, 58

[243] Thang Duy Dao, Kai Chen, Satoshi Ishii, Akihiko Ohi, Toshihide Nabatame, Masahiro Kitajima, and Tadaaki Nagao. Infrared perfect absorbers fabricated by colloidal mask etching of Al–Al$_2$O$_3$ –Al trilayers. *ACS Photonics*, 2(7):964–970, Jul 2015. 57

[244] Hu Tao, C M Bingham, D Pilon, Kebin Fan, A C Strikwerda, D Shrekenhamer, W J Padilla, Xin Zhang, and R D Averitt. A dual band terahertz metamaterial absorber. *J. Phys. D. Appl. Phys.*, 43(22):225102, Jun 2010. DOI: 10.1088/0022-3727/43/22/225102 57

[245] Xinyu Liu and Willie J. Padilla. Thermochromic infrared metamaterials. *Adv. Mater.*, 28(5):871–875, Feb 2016. DOI: 10.1002/adma.201504525 57, 58

[246] Qi Xiao, Timothy U. Connell, Jasper J. Cadusch, Ann Roberts, Anthony S. R. Chesman, and Daniel E. Gómez. Hot-carrier organic synthesis via the near-perfect absorption of light. *ACS Catal.*, 8(11):10331–10339, Nov 2018. DOI: 10.1021/acscatal.8b03486 59

[247] Corey Shemelya, Dante DeMeo, Nicole Pfiester Latham, Xueyuan Wu, Chris Bingham, Willie Padilla, and Thomas E. Vandervelde. Stable high temperature metamaterial emitters for thermophotovoltaic applications. *Applied Physics Letters*, 104(20):201113, May 2014. DOI: 10.1063/1.4878849 59

[248] Zhongyang Li, Edgar Palacios, Serkan Butun, Hasan Kocer, and Koray Aydin. Omnidirectional, broadband light absorption using large-area, ultrathin lossy metallic film coatings. *Sci. Rep.*, 5(1):15137, Dec 2015. DOI: 10.1038/srep15137 57

[249] Huixu Deng, Zhigang Li, Liliana Stan, Daniel Rosenmann, David Czaplewski, Jie Gao, and Xiaodong Yang. Broadband perfect absorber based on one ultrathin layer of refractory metal. *Opt. Lett.*, 40(11):2592, Jun 2015. DOI: 10.1364/ol.40.002592 57

[250] Prakash Pitchappa, Chong Pei Ho, Piotr Kropelnicki, Navab Singh, Dim-Lee Kwong, and Chengkuo Lee. Dual band complementary metamaterial absorber in near infrared region. *J. Appl. Phys.*, 115(19):193109, May 2014. DOI: 10.1063/1.4878459 57, 58

[251] Wei Wang, Yurui Qu, Kaikai Du, Songang Bai, Jingyi Tian, Meiyan Pan, Hui Ye, Min Qiu, and Qiang Li. Broadband optical absorption based on single-sized metal-dielectric-metal plasmonic nanostructures with high- ϵ metals. *Appl. Phys. Lett.*, 110(10):101101, Mar 2017. DOI: 10.1063/1.4977860 59

[252] Fei Ding, Jin Dai, Yiting Chen, Jianfei Zhu, Yi Jin, and Sergey I. Bozhevolnyi. Broadband near-infrared metamaterial absorbers utilizing highly lossy metals. *Sci. Rep.*, 6(1):39445, Dec 2016. DOI: 10.1038/srep39445 57

[253] Lei Lei, Shun Li, Haixuan Huang, Keyu Tao, and Ping Xu. Ultra-broadband absorber from visible to near-infrared using plasmonic metamaterial. *Opt. Express*, 26(5):5686, Mar 2018. DOI: 10.1364/oe.26.005686 57

[254] Yijia Huang, Ling Liu, Mingbo Pu, Xiong Li, Xiaoliang Ma, and Xiangang Luo. A refractory metamaterial absorber for ultra-broadband, omnidirectional and polarization-independent absorption in the UV-NIR spectrum. *Nanoscale*, 10(17):8298–8303, May 2018. DOI: 10.1039/c8nr01728j 59

[255] Chun-Chieh Chang, Wilton J. M. Kort-Kamp, John Nogan, Ting S. Luk, Abul K. Azad, Antoinette J. Taylor, Diego A. R. Dalvit, Milan Sykora, and Hou-Tong Chen. High-temperature refractory metasurfaces for solar thermophotovoltaic energy harvesting. *Nano Letters*, 18(12):7665–7673, Dec 2018. DOI: 10.1021/acs.nanolett.8b03322 59

[256] Hyung Ki Kim, Dongju Lee, and Sungjoon Lim. Wideband-switchable metamaterial absorber using injected liquid metal. *Sci. Rep.*, 6(1):31823, Oct 2016. DOI: 10.1038/srep31823 57

[257] Cheng Zhang, Qiang Cheng, Jin Yang, Jie Zhao, and Tie Jun Cui. Broadband metamaterial for optical transparency and microwave absorption. *Appl. Phys. Lett.*, 110(14):143511, Apr 2017. DOI: 10.1063/1.4979543 57

[258] Wei Li, Urcan Guler, Nathaniel Kinsey, Gururaj V. Naik, Alexandra Boltasseva, Jianguo Guan, Vladimir M. Shalaev, and Alexander V. Kildishev. Refractory plasmonics with titanium nitride: broadband metamaterial absorber. *Adv. Mater.*, 26(47):7959–7965, Dec 2014. DOI: 10.1002/adma.201401874 57

[259] Lujun Huang, Guoqing Li, Alper Gurarslan, Yiling Yu, Ronny Kirste, Wei Guo, Junjie Zhao, Ramon Collazo, Zlatko Sitar, Gregory N. Parsons, Michael Kudenov, and Linyou Cao. Atomically thin MoS$_2$ narrowband and broadband light superabsorbers. *ACS Nano*, 10(8):7493–7499, Aug 2016. 57

[260] ChengGang Hu, Xiong Li, Qin Feng, Xu'Nan Chen, and XianGang Luo. Investigation on the role of the dielectric loss in metamaterial absorber. *Opt. Express*, 18(7):6598, Mar 2010. DOI: 10.1364/oe.18.006598 58

[261] Wangren Xu and Sameer Sonkusale. Microwave diode switchable metamaterial reflector/absorber. *Appl. Phys. Lett.*, 103(3):031902, Jul 2013. DOI: 10.1063/1.4813750 58

[262] H. Yuan, B. O. Zhu, and Y. Feng. A frequency and bandwidth tunable metamaterial absorber in x-band. *J. Appl. Phys.*, 117(17):173103, May 2015. DOI: 10.1063/1.4919753 58

[263] Yunsong Xie, Xin Fan, Jeffrey D. Wilson, Rainee N. Simons, Yunpeng Chen, and John Q. Xiao. A universal electromagnetic energy conversion adapter based on a metamaterial absorber. *Sci. Rep.*, 4(1):6301, May 2015. DOI: 10.1038/srep06301 58

[264] Young Joon Yoo, Sanghyun Ju, Sang Yoon Park, Young Ju Kim, Jihye Bong, Taekyung Lim, Ki Won Kim, Joo Yull Rhee, and YoungPak Lee. Metamaterial absorber for electromagnetic waves in periodic water droplets. *Sci. Rep.*, 5(1):14018, Nov 2015. DOI: 10.1038/srep14018 58

[265] Yongqiang Pang, Jiafu Wang, Qiang Cheng, Song Xia, Xiao Yang Zhou, Zhuo Xu, Tie Jun Cui, and Shaobo Qu. Thermally tunable water-substrate broadband metamaterial absorbers. *Appl. Phys. Lett.*, 110(10):104103, Mar 2017. DOI: 10.1063/1.4978205 58

[266] Qinghua Song, Wu Zhang, Pin Chieh Wu, Weiming Zhu, Zhong Xiang Shen, Peter Han Joo Chong, Qing Xuan Liang, Zhen Chuan Yang, Yi Long Hao, Hong Cai, Hai Feng Zhou, Yuandong Gu, Guo-Qiang Lo, Din Ping Tsai, Tarik Bourouina, Yamin

Leprince-Wang, and Ai-Qun Liu. Water-resonator-based metasurface: an ultrabroadband and near-unity absorption. *Adv. Opt. Mater.*, 5(8):1601103, Apr 2017. DOI: 10.1002/adom.201601103 58

[267] Chong He, FaJun Xiao, Ivan D. Rukhlenko, Jianwen Xie, Junping Geng, Malin Premaratne, Ronghong Jin, Weiren Zhu, and Xianling Liang. Water metamaterial for ultra-broadband and wide-angle absorption. *Opt. Express, Vol. 26, Issue 4, pp. 5052-5059*, 26(4):5052–5059, Feb 2018. DOI: 10.1364/oe.26.005052 58

[268] Junming Zhao, Shu Wei, Cheng Wang, Ke Chen, Bo Zhu, Tian Jiang, and YiJun Feng. Broadband microwave absorption utilizing water-based metamaterial structures. *Opt. Express*, 26(7):8522, Apr 2018. DOI: 10.1364/oe.26.008522 58

[269] Yang Shen, Jieqiu Zhang, Yongqiang Pang, Jiafu Wang, Hua Ma, and Shaobo Qu. Transparent broadband metamaterial absorber enhanced by water-substrate incorporation. *Opt. Express*, 26(12):15665, Jun 2018. DOI: 10.1364/oe.26.015665 58

[270] Y. J. Yoo, H. Y. Zheng, Y. J. Kim, J. Y. Rhee, J.-H. Kang, K. W. Kim, H. Cheong, Y. H. Kim, and Y. P. Lee. Flexible and elastic metamaterial absorber for low frequency, based on small-size unit cell. *Appl. Phys. Lett.*, 105(4):041902, Jul 2014. DOI: 10.1063/1.4885095 59

[271] Taehee Jang, Hongseok Youn, Young Jae Shin, and L. Jay Guo. Transparent and flexible polarization-independent microwave broadband absorber. *ACS Photonics*, 1(3):279–284, Mar 2014. DOI: 10.1021/ph400172u 57, 59

[272] Longqing Cong, Siyu Tan, Riad Yahiaoui, Fengping Yan, Weili Zhang, and RanJan Singh. Experimental demonstration of ultrasensitive sensing with terahertz metamaterial absorbers: A comparison with the metasurfaces. *Appl. Phys. Lett.*, 106(3):031107, Jan 2015. DOI: 10.1063/1.4906109 58

[273] Ziqi Miao, Qiong Wu, Xin Li, Qiong He, Kun Ding, Zhenghua An, Yuanbo Zhang, and Lei Zhou. Widely tunable terahertz phase modulation with gate-controlled graphene metasurfaces. *Phys. Rev. X*, 5(4):041027, Nov 2015. DOI: 10.1103/physrevx.5.041027 58

[274] Xiaoguang Zhao, Jacob Schalch, Jingdi Zhang, Huseyin R. Seren, Guangwu Duan, Richard D. Averitt, and Xin Zhang. Electromechanically tunable metasurface transmission waveplate at terahertz frequencies. *Optica*, 5(3):303, Mar 2018. DOI: 10.1364/optica.5.000303 58

[275] Guangwu Duan, Jacob Schalch, Xiaoguang Zhao, Jingdi Zhang, Richard D. Averitt, and Xin Zhang. An air-spaced terahertz metamaterial perfect absorber. *Sensors Actuators A Phys.*, 280:303–308, Sep 2018. 58

[276] Fabio Alves, Brian Kearney, Dragoslav Grbovic, and Gamani Karunasiri. Narrowband terahertz emitters using metamaterial films. *Opt. Express*, 20(19):21025, Sep 2012. DOI: 10.1364/oe.20.021025 58

[277] Xin Hu, Gaiqi Xu, Long Wen, Huacun Wang, Yuncheng Zhao, Yaxin Zhang, David R. S. Cumming, and Qin Chen. Metamaterial absorber integrated microfluidic terahertz sensors. *Laser Photon. Rev.*, 10(6):962–969, Nov 2016. DOI: 10.1002/lpor.201600064 58

[278] Goran Isić, Borislav Vasić, Dimitrios C. Zografopoulos, Romeo Beccherelli, and Radoš Gajić. Electrically tunable critically coupled terahertz metamaterial absorber based on nematic liquid crystals. *Phys. Rev. Appl.*, 3(6):064007, Jun 2015. DOI: 10.1103/physrevapplied.3.064007 59

[279] Shichao Song, Qin Chen, Lin Jin, and Fuhe Sun. Great light absorption enhancement in a graphene photodetector integrated with a metamaterial perfect absorber. *Nanoscale*, 5(20):9615, Sep 2013. DOI: 10.1039/c3nr03505k 59

[280] Junghyun Park, Ju-Hyung Kang, Xiaoge Liu, and Mark L. Brongersma. Electrically tunable epsilon-near-zero (ENZ) metafilm absorbers. *Sci. Rep.*, 5(1):15754, Dec 2015. DOI: 10.1038/srep15754 59

[281] Yuki Matsuno and Atsushi Sakurai. Perfect infrared absorber and emitter based on a large-area metasurface. *Opt. Mater. Express*, 7(2):618, Feb 2017. DOI: 10.1364/ome.7.000618 59

[282] H. T. Miyazaki, T. Kasaya, M. Iwanaga, B. Choi, Y. Sugimoto, and K. Sakoda. Dual-band infrared metasurface thermal emitter for CO_2 sensing. *Appl. Phys. Lett.*, 105(12):121107, Sep 2014. 58

[283] Fei Yi, Euijae Shim, Alexander Y. Zhu, Hai Zhu, Jason C Reed, and Ertugrul Cubukcu. Voltage tuning of plasmonic absorbers by indium tin oxide. *Appl. Phys. Lett.*, 102(22):221102, Jun 2013. DOI: 10.1063/1.4809516 58

[284] D. Costantini, A. Lefebvre, A.-L. Coutrot, I. Moldovan-Doyen, J.-P. Hugonin, S. Boutami, F. Marquier, H. Benisty, and J.-J. Greffet. Plasmonic metasurface for directional and frequency-selective thermal emission. *Phys. Rev. Appl.*, 4(1):014023, Jul 2015. DOI: 10.1103/physrevapplied.4.014023 58

[285] Ramon Walter, Andreas Tittl, Audrey Berrier, Florian Sterl, Thomas Weiss, and Harald Giessen. Large-area low-cost tunable plasmonic perfect absorber in the near infrared by colloidal etching lithography. *Adv. Opt. Mater.*, 3(3):398–403, Mar 2015. DOI: 10.1002/adom.201400545 58

[286] Md Muntasir Hossain, Baohua Jia, and Min Gu. A metamaterial emitter for highly efficient radiative cooling. *Adv. Opt. Mater.*, 3(8):1047–1051, Aug 2015. DOI: 10.1002/adom.201500119 58

[287] Peng Yu, Lucas V. Besteiro, Jiang Wu, Yongjun Huang, Yueqi Wang, Alexander O. Govorov, and Zhiming Wang. Metamaterial perfect absorber with unabated size-independent absorption. *Opt. Express*, 26(16):20471, Aug 2018. DOI: 10.1364/oe.26.020471 58

[288] Yu Hui, Juan Sebastian Gomez-Diaz, Zhenyun Qian, Andrea Alù, and Matteo Rinaldi. Plasmonic piezoelectric nanomechanical resonator for spectrally selective infrared sensing. *Nat. Commun.*, 7(1):11249, Dec 2016. DOI: 10.1038/ncomms11249 58

[289] Khagendra Bhattarai, Zahyun Ku, Sinhara Silva, Jiyeon Jeon, Jun Oh Kim, Sang Jun Lee, Augustine Urbas, and Jiangfeng Zhou. A large-area, mushroom-capped plasmonic perfect absorber: refractive index sensing and Fabry-Perot cavity mechanism. *Adv. Opt. Mater.*, 3(12):1779–1786, Dec 2015. DOI: 10.1002/adom.201500231 60

[290] G. X. Li, S. M. Chen, W. H. Wong, E. Y. B. Pun, and K. W. Cheah. Highly flexible near-infrared metamaterials. *Opt. Express*, 20(1):397, Jan 2012. DOI: 10.1364/oe.20.000397 58

[291] Kaikai Du, Lu Cai, Hao Luo, Yue Lu, Jingyi Tian, Yurui Qu, Pintu Ghosh, Yanbiao Lyu, Zhiyuan Cheng, Min Qiu, and Qiang Li. Wavelength-tunable mid-infrared thermal emitters with a non-volatile phase changing material. *Nanoscale*, 10(9):4415–4420, Mar 2018. DOI: 10.1039/c7nr09672k 58

[292] T. U. Tumkur, Lei Gu, J. K. Kitur, E. E. Narimanov, and M. A. Noginov. Control of absorption with hyperbolic metamaterials. *Appl. Phys. Lett.*, 100(16):161103, Apr 2012. DOI: 10.1063/1.4703931 58

[293] Charlene Ng, Jasper J. Cadusch, Svetlana Dligatch, Ann Roberts, Timothy J. Davis, Paul Mulvaney, and Daniel E. Gómez. Hot carrier extraction with plasmonic broadband absorbers. *ACS Nano*, 10(4):4704–4711, Apr 2016. DOI: 10.1021/acsnano.6b01108 59

[294] David N. Woolf, Emil A. Kadlec, Don Bethke, Albert D. Grine, John J. Nogan, Jeffrey G. Cederberg, D. Bruce Burckel, Ting Shan Luk, Eric A. Shaner, and Joel M. Hensley. High-efficiency thermophotovoltaic energy conversion enabled by a metamaterial selective emitter. *Optica*, 5(2):213, Feb 2018. DOI: 10.1364/optica.5.000213 59

[295] Ward D. Newman, Cristian L. Cortes, Jon Atkinson, Sandipan Pramanik, Raymond G. DeCorby, and Zubin Jacob. Ferrell–berreman modes in plasmonic epsilon-near-zero media. *ACS Photonics*, 2(1):2–7, December 2014. DOI: 10.1021/ph5003297 60

[296] Salvatore Campione, Iltai Kim, Domenico de Ceglia, Gordon A. Keeler, and Ting S. Luk. Experimental verification of epsilon-near-zero plasmon polariton modes in degenerately doped semiconductor nanolayers. *Optics Express*, 24(16):18782, August 2016. DOI: 10.1364/oe.24.018782 60

[297] Simon Vassant, Jean-Paul Hugonin, Francois Marquier, and Jean-Jacques Greffet. Berreman mode and epsilon near zero mode. *Optics Express*, 20(21):23971, October 2012. DOI: 10.1364/oe.20.023971 60

[298] S. Vassant, A. Archambault, F. Marquier, F. Pardo, U. Gennser, A. Cavanna, J. L. Pelouard, and J. J. Greffet. Epsilon-near-zero mode for active optoelectronic devices. *Physical Review Letters*, 109(23), December 2012. DOI: 10.1103/physrevlett.109.237401 60

[299] Aleksei Anopchenko, Long Tao, Catherine Arndt, and Ho Wai Howard Lee. Field-effect tunable and broadband epsilon-near-zero perfect absorbers with deep subwavelength thickness. *ACS Photonics*, 5(7):2631–2637, Jul 2018. DOI: 10.1021/acsphotonics.7b01373 60

[300] Yuanmu Yang, Kyle Kelley, Edward Sachet, Salvatore Campione, Ting S. Luk, Jon-Paul Maria, Michael B. Sinclair, and Igal Brener. Femtosecond optical polarization switching using a cadmium oxide-based perfect absorber. *Nat. Photonics*, 11(6):390–395, Jun 2017. DOI: 10.1038/nphoton.2017.64 60

[301] Xinxiang Niu, Xiaoyong Hu, Saisai Chu, and Qihuang Gong. Epsilon-near-zero photonics: a new platform for integrated devices. *Advanced Optical Materials*, 6(10):1701292, March 2018. DOI: 10.1002/adom.201701292 60

[302] Nathaniel Kinsey, Clayton DeVault, Alexandra Boltasseva, and Vladimir M. Shalaev. Near-zero-index materials for photonics. *Nature Reviews Materials*, 4(12):742–760, September 2019. DOI: 10.1038/s41578-019-0133-0 60

[303] Shulin Sun, Qiong He, Shiyi Xiao, Qin Xu, Xin Li, and Lei Zhou. Gradient-index meta-surfaces as a bridge linking propagating waves and surface waves. *Nature Materials*, 11(5):426–431, April 2012. DOI: 10.1038/nmat3292 60

[304] Jianxiong Li, Ping Yu, Chengchun Tang, Hua Cheng, Junjie Li, Shuqi Chen, and Jianguo Tian. Bidirectional perfect absorber using free substrate plasmonic metasurfaces. *Advanced Optical Materials*, 5(12):1700152, May 2017. DOI: 10.1002/adom.201700152 60

[305] Wen-Chen Chen, Andrew Cardin, Machhindra Koirala, Xianliang Liu, Talmage Tyler, Kevin G. West, Christopher M. Bingham, Tatiana Starr, Anthony F. Starr, Nan M.

Jokerst, and Willie J. Padilla. Role of surface electromagnetic waves in metamaterial absorbers. *Optics Express*, 24(6):6783, March 2016. DOI: 10.1364/oe.24.006783 60

CHAPTER 4

Fabrication of Metamaterial Perfect Absorbers

Over the past two decades, the development of metamaterials benefited from the progression of state-of-the-art fabricational techniques. Electrodynamic similitude [1] indicates that metamaterials properties may be obtained at nearly any wavelength by a simple scaling of their dimensions. However, the realization of scaled designs is increasingly challenging as the operational spectral range shifts to shorter wavelength, i.e., in visible and ultraviolet regions. In this chapter, we will give a comprehensive overview of various fabrication techniques, which are used to obtain metamaterial perfect absorbers working in the spectrum from terahertz through infrared to visible. Their features, as well as associated challenges and issues, will also be discussed.

4.1   INTRODUCTION

Electromagnetic metamaterials are composite artificial electromagnetic materials, whose electromagnetic properties are determined by the geometry of the subwavelength meta-atoms (unit cells) rather than by the particular properties of their constituents. Over the past two decades, metamaterials have demonstrated unprecedented properties that are not easily attained by naturally occurring materials, and have shown these unique responses over much of the electromagnetic spectrum, from radio frequency through terahertz to the ultraviolet. Since inception, metamaterials have been shown at increasingly smaller wavelengths, and thus feature sizes have shrunk progressively from centimeter to micrometers and the nanometer scale-enabled from the maturity of fabricational techniques in micro-/nano-electronics. The associated methods allow us to experimentally realize rationally engineered scattering that arise from the interaction of electromagnetic waves with metamaterials. Conventional photo-lithographic techniques with flooding UV light on a planar substrate, coated with photosensitive resist, which were first invented for the integrated circuit industry, have been widely adopted for metamaterial patterning. In addition to typical photo-lithography, electron-beam lithography (EBL) allows for direct writing of patterns on a electron-beam sensitive resist in a raster-scanning manner, and is also frequently used to fabricate metamaterials with feature sizes as small as tens of nanometers. With these techniques, metamaterials have shown exotic electromagnetic responses including invisible cloaking [2], negative refractive index [3, 4], perfect absorption [5–8], and Huygens' meta-atoms [9, 10].

In addition to the more common fabricational techniques mentioned above, there are also many other alternative techniques, for example focus-ion-beam (FIB) lithography, and interference lithography. We overview micro-/nano-fabricational techniques and categorize them into two major branches: masked fabrication and maskless fabrication. However, our review of fabricational methods will not be limited only to metamaterials absorbers, but will also include more general metamaterial fabrication, since the techniques are general, and can be applied. Masked fabricational techniques produce metamaterial patterns via applying masks or molds over a substrate, such as conventional photo-lithography, nanoimprint lithography, and shadow mask lithography. Maskless fabricational techniques usually lead to patterns without the aid of a specific mask, such as electron-beam lithography, interference lithography, self-assembly technique, and multi-photon lithography. We will also discuss the major associated challenges and issues of these techniques. Since we focus on the lithography techniques for patterning metamaterials, other related techniques, including evaporation, deposition, and etching, will not be addressed here. Further information on these topics can be found in Ref. [11].

4.2   MASKED LITHOGRAPHY

4.2.1   PHOTOLITHOGRAPHY

Photolithography is a mask-assisted technique which was invented initially by Nicephore Niepce in the early 19th century for photography using lithographic stones [12]. Then in 1950s, this technique was adopted to generate small-size electronic circuits. A UV light source (usually monochromatic) is used to insolate a photosensitive polymer layer, called photoresist, through a mask on which the patterns are projected via lenses on to a sample. Modern photolithographic techniques have become a key technique for the production of Very Large-Scale Integration (VLSI) of electronics.

Due to the many developed standard process recipes and low cost for large-scale production, photolithographic techniques are one of the most common methods used to synthesize metamaterials on scales from microns to millimeters, where critical dimensions are limited by the wavelength of UV source due to the diffraction limit. Using a stepper combined with a deep UV (DUV) source, of KrF laser with wavelength of 248 nm, the critical dimensions of approximately 100 nm can be achieved. After photolithography, mask patterns are transferred to a photosensitive resist after the development process. It is then possible to obtain printed patterns on a substrate via further etching or material deposition process. For metal-type metamaterials, after deposition, a lift-off process through dissolving the undeveloped resist in solvents may be followed. Figures 3.4b–c show SEM pictures of fabricated metallic metamaterials, which exhibit resonances at mid-infrared and near-infrared wavelengths, respectively.

4.2.2   SHADOW MASK LITHOGRAPHY

Shadow mask lithography (SML), also as known as stencil lithography, is a resist-free and liquid-free fabrication technique involving a direct deposition of metal, and dielectric thin films onto a substrate which is masked, using a perforated stencil membrane. To implement SML, a micro- or nanoscale stencil mask, which is normally made of thin films, such as low-stress silicon nitride [13, 14], and silicon membranes [15], is initially prepared. The stencil is then placed either in direct contact or in close proximity to the target substrate. To transfer the patterns from the stencil to the substrate, a following metal or dielectric deposition process is performed. Figure 4.1a shows a typical SML sequence including the process to make a stencil mask out of a thin layer of silicon nitride based on a traditional micromachining technique [13]. The stencil masks can be further reused to replicate the patterns with repeating the process. Using SML, feature sizes as small as 100 nm can be achieved.

Since there is no chemical or heat treatment involved, the SML technique has been widely used for patterning planar micro- and nano-scale structures onto substrates sensitive to chemical degradation or resist contamination, such as strongly correlated materials [16] and graphene [17]. To protect the phase change material, vanadium dioxide (VO_2), from chemical erosion [18, 19], Liu et al. utilized a SiN_x micro-stencil to pattern metamaterials on an intact VO_2 thin film, in order to drive dynamics which leads to insulator-to-metal phase transition, enhancing the sensitivity to the local conductivity changes [16]. Figure 4.1b shows a microscopic image of the fabricated metamaterial array, in gaps of which the VO_2 thin film is damaged due to the significantly enhanced carrier dynamics. It can be observed that the metamaterial gaps are reduced from 8 μm to about 2 μm because of shadow diffusion resulting from the spacing between the stencil mask and rigid substrate. To mitigate the blurring of patterns, applying stencil masks onto flexible substrates would allow very close contact. Figure 4.1c illustrates patterned arrayed antennas on a flexible PDMS substrate. Compared to the patterning on a silicon substrate, the resolution is much better on the flexible PDMS. With nearly zero-gap contact, a sub-50-nm gap feature can be obtained, as shown in Fig. 4.1d [14].

Although shadow mask lithography has shown its potential as a resistless, scalable, and reusable technique for large-area and low-cost fabrication on a variety of substrates, there are still several limitations and challenges that need to be overcome for its further development. The major limitation for the SML nano-structuring technique lies in the patterning of continuous structures, such as meander lines and Babinet metamaterials [20, 21], where most area of the substrate is covered by deposited materials, e.g., gold. For example, to produce Babinet metamaterials using SML, patterns on the stencil mask have to be isolated from each other without interconnections, and thus it is impossible to achieve this on a free-standing stencil membrane. Perforated stencil masks made of free-standing sub-μm thick silicon nitride or silicon membrane are quite stable for repetitive use, it is nonetheless challenging to protect them from breaking during contact, due to the uneven substrate surface or residue particles left on the substrate surface. Therefore, development of polymeric stamps capable of patterning micro-

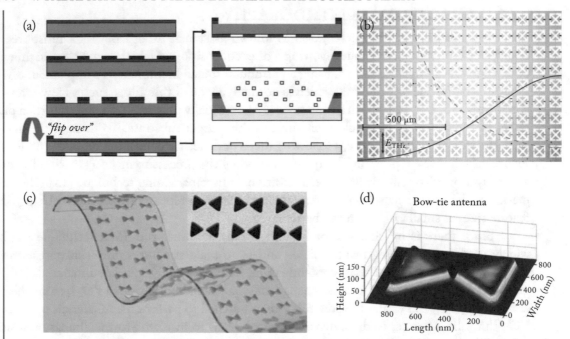

Figure 4.1: (a) Illustration of fabrication process of THz metamaterial on silk using SML technique. (b) Microscopic image of the patterned metamaterial array on a vanadium dioxide (VO_2) substrate. (c) Illustration of fabricated bow-tie antennas on a flexible PDMS substrate using SML technique. The inset shows an SEM image of the fabricated stencil mask on a SiN_x membrane. (d) An AFM image of a bow-tie particle with sharp edges and a sub-50-nm gap distance on a flexible substrate. Reprinted with permission from: (a) Ref. [13] © 2010 Wiley-VCH; (b) Ref. [16] © 2012 NPG; (c) and (d) Ref. [14] © 2011 Wiley-VCH.

and nano-structures would impact the potential of SML for industrial applications. Another challenge for SML is the deterioration of the geometry of patterns due to the accumulation of the deposited materials around the peripheral of the apertures. This is more severe for aperture sizes down to sub-microns [14]. Therefore, a selective etching process that only removes accumulating materials, while keeping the stencil geometry intact, should be developed. With growing interest in the shadow mask lithographic technique, efforts have been made to tackle these challenges with novel methods, such as compliant stencil mask on cantilevers [22], and natural stencil masks made of diatoms [23]. Shadow mask lithography, with its great advantages for low-cost, high-speed, easy-scaling, and chemical-free fabrication, will be an important addition to the conventional fabrication technique for metamaterial structuring.

4.2.3 NANOIMPRINT LITHOGRAPHY (NIL)

Since the initial work by Chou et al. using a SiO_2 mold to pattern sub-50-nm structures [24, 25], nanoimprint lithography (NIL) has been widely recognized as a potential low-cost, high-speed, and high-throughput method for micro- and nanostructuring. Due to these advantages, NIL was added on to the ITRS (International Technology Roadmap for Semiconductors) as a candidate technology for the 32 and 22-nm nodes in 2003. Because of its simplicity, this technique has been widely used in various areas other than electronics, such as biomedical applications, micro- and nanofluidics, photonics, and physical sensors [26, 27]. In recent years, it has drawn much attention within the metamaterial community and several examples have been demonstrated in the visible and infrared ranges [28–32].

The basic principle of NIL is based on mechanical deformation of resist. In the NIL process, a prefabricated hard mold that contains micro- and nanoscale inverse features—compared to the original designs—is mechanically pressed onto a resist-coated substrate to allow the resist layer squeeze into cavities of the mold to form arbitrary shapes after curing. The created structures can achieve resolution down to sub-10-nm [33, 34], which is beyond the diffraction limit of light as observed in conventional lithography techniques. The mold material used in NIL should have a very large contrast of Young's Modulus compared to the resist layer. Candidate materials include Si, SiO_2, SiC, silicon nitride, metals, and polymeric materials. The resist material should have a much lower viscosity to be easily deformed under pressure and maintain a sufficient mechanical strength and good mold-releasing properties during de-molding process after resist curing [26]. In terms of resist curing process, there are two major types for NIL: thermal NIL and UV NIL. Compared to thermal NIL, in which high temperature above the glass transition temperature (T_g) is required to melt the resist, UV NIL, where the resist is cured by UV exposure, operates in a short period at room temperature without heating and cooling process [27].

In 2005, room-temperature NIL was successfully applied for patterning metallic and dielectric planar chiral metamaterials with linewidth around 100-nm [28]. The mode patterns were made from silicon and fabricated by EBL and RIE process. The metallic structures were obtained by fabricating a bilayer structure consisting of a 150-nm thick PMMA on the bottom and a 30-nm-thick hydrogen silsequioxane (HSQ) on the top. The NIL process was directly carried out on the top HSQ layer via pressing the rigid flat mold onto a rigid flat substrate, i.e., plate-to-plate NIL. After that, the patterns on the HSQ layer were transferred to the bottom the PMMA layer via O_2 etching. The final metal layer was finished by lift-off process after metal deposition. Figures 4.2a–b display the SEM images of the fabricated chiral photonic metamaterials on a silicon substrate. Using similar method, Wu et al. demonstrated fishnet metamaterials exhibiting negative refractive index operating at near-IR and mid-IR wavelengths, respectively.

In order to create a large-area pattern through the plate-to-plate NIL process, a highly accurate alignment during imprinting is necessary to repeat the patterns on the mold step-by-step. As a result, this method leads to longer operational time and system complexity [35]. To fur-

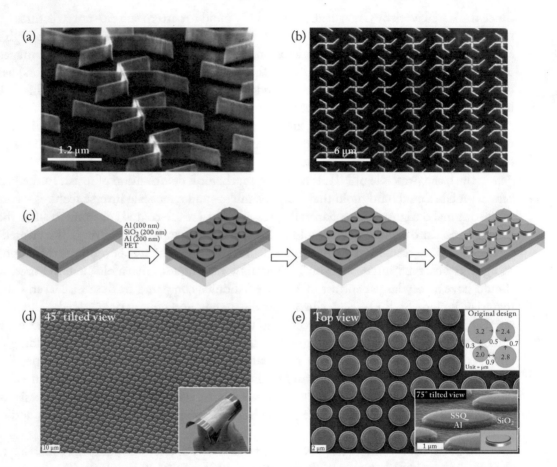

Figure 4.2: (a) SEM image of silicon mode structure after over 50-time NIL process. (b) SEM image of the metallic planar chiral metamaterials fabricated via NIL process. (c) Illustration of a roll-to-roll (R2R) NIL process to fabricate metal-insulation-metal based plasmonic metamaterials on a flexible PET substrate, which is coated with Al/SiO$_2$/Al thin films. The inset shows an SEM image of the fabricated stencil mask on a SiN$_x$ membrane. (d) SEM image of the fabricated plasmonic metamaterial film comprising the Al disk array via the R2R NIL process on SiO2/Al/PET substrates. The scalability and flexibility are exemplified in the inset to (d). The enlarged top view (e) discloses the fabricated disk pattern geometry that is almost identical to the original design described in the upper-right corner. The inset reveals the remaining SSQ mask on top of Al disk patches. Reprinted with permission from: (a) and (b) Ref. [28] © 2005 Elsevier; (c)–(e) Ref. [32] © 2012 AIP.

ther achieve a scalable and much faster imprinting process, a roll-to-roll (R2R) NIL technique capable of large-area fabrication of micro/nanoscale structures was introduced in 2006 [36]. In this process, an imprinted roller with a patterned surface or wrapped with a flexible mold is used to imprint onto a flexible substrate supported by another roller, thus enabling continuous fabrication process with a high throughput. In 2012, this method was first employed by Guo et al. to demonstrate large-area flexible metamaterials. Figure 4.2c illustrates the R2R NIL fabrication process on a PET substrate coated with $Al/SiO_2/Al$ thin films. The imprinting mold consisted of a flexible PDMS substrate with patterns replicated from a 4-inch wafer-scale Si pillar master. During imprint process, the PDMS pad was wrapped around a roller in the 6-inch compatible R2R NIL apparatus [37]. A UV-curable epoxy-silsesquioxane (SSQ) resist layer was used to transfer the patterns on the mold to the PET substrate. Compared to other lift-off processes used to define metamaterials, a subsequent dry/wet etching process completes the final structure with various Al disks, as shown in Figs. 4.2d–e.

Large-area production of functional optical and infrared metamaterials has been desired for a long time. Due to the high-resolution and high-throughput process, NIL is well suited for wafer-scale metamaterial production with feature size down to sub-microns. However, there are still some challenges which must be overcome in order for NIL to produce large area nano/micro-scale metamaterials. First, it should be noted that the resolution of the process is limited by the resolution of molds, which are normally fabricated using other techniques, such as UV lithography for micron-size features, EBL for nanoscale features, interference lithography for large-area periodic structures. Second, to cure the resist using ultraviolet, either the substrate or the mold is required to be transparent to an appropriate UV source. For most metallic perfect absorber designs, the opaque ground plane further limits the selection of mold materials.

4.3 MASKLESS LITHOGRAPHY

4.3.1 ELECTRON-BEAM LITHOGRAPHY (EBL)

Electron-beam lithography (EBL) is an alternative and widely used technique to fabricate planar metamaterials with feature size as small as 100 nm. In comparison to DUV stepper techniques, EBL allows for patterning nearly arbitrary 2D patterns with a resolution down to 10-nm directly on a substrate, without requiring an expensive mask made ahead of sample fabrication. Therefore, this flexible technique is very favorable for iteration of preliminary metamaterial designs before mass production, and is capable of achieving structures with critical dimensions not easily attainable from other techniques.

Analogous to conventional photolithography techniques, EBL performs exposure on an electron-beam resist, including positive-tone resist-like PMMA, ZEP, and negative tone resist like maN2400, $H-SiO_x$ (HSQ), by a high-energy electron beam raster-scanned over the sample directly via a computer generated pattern rather than a mask for pattern transferring. Subsequently, the etching/deposition and liftoff process will be performed after developing the exposed resist. Because of the high-resolution ability to create nanoscale features, EBL has be-

(a) 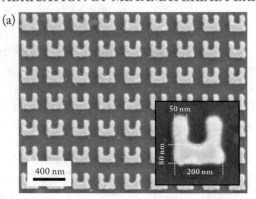 (b)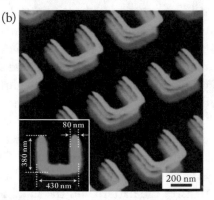

Figure 4.3: (a) SEM image of an SRR array with a total size of $(100 \ \mu m)^2$, exhibiting resonances at telecommunication and visible frequencies. The inset shows the dimensions of the unit-cell element. (b) SEM image of a four-layer SRR array with an oblique view. The inset shows the dimensions of a unit cell. Reprinted with permission from: (a) Ref. [39] © 2005 APS; (b) Ref. [40] © 2008 NPG.

come a powerful technique to fabricate subwavelength metamaterial resonators at infrared and visible wavelengths. Metamaterials fabricated from EBL processes were initially demonstrated in the mid-infrared and telecommunication wavelengths [38, 39]. Figure 4.3a shows the SEM image of an SRR array fabricated on a glass substrate. The total size of the metallic resonator is about 200 nm while the smallest feature size is only about 50 nm, both of which are much smaller than the operational wavelength of 1.5 μm.

In addition to patterning a single layer of planar metamaterial structures, EBL can also be employed for constructing multiple-layer metamaterial composites with multiple EBL runs using registration of adjacent layers. Between the successive lithography processes, an intermediate planarization step can be implemented to make sure that the electron beam is focused on the resist layer for better sharpness of exposure [40, 41]. Figure 4.3b shows the first four-layer metamaterial stack with feature size around 80-nm [40]. The inter-layer metamaterials are embedded in a solidifiable photopolymer PC403, which was also used as a planarization layer in the process. The misalignment between the adjacent layers can be well controlled within 10 nm. Such an EBL-multilayer technique was further used to demonstrate an infrared metamaterial phase hologram consisting of three layers of gold elements in a SiO_2 matrix [41].

While EBL is a compelling technique to achieve structures down to tens of nanometers, it is limited in creating large-area metamaterials due to the serial nature of writing, i.e., from point to point or line to line. Such a serial process may lead to an extremely long writing time that causes potential issues of beam instability and astigmatism. Moreover, tessellating numerous small cells together to obtain a large size pattern introduces stitching errors, which could dislocate structures between adjacent cells.

(a)  (b)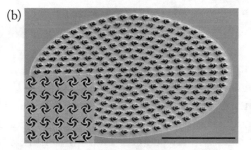

Figure 4.4: (a) SEM image of an FIB fabricated 3D fishnet negative-index metamaterial prism with a unit cell periodicity of 860-nm, height of 800-nm, and feature size of 265-nm. The inset shows a magnified view with the layers visible in each hole. (b) An oblique-view SEM image of an FIB fabricated circular grating consisting of left-handed (LH) and right-handed (RH) pinwheel structures. The inset shows the SEM image of a linear grating consisting of both LH and RH pinwheels, which were carved out of a 80-nm free-standing gold film. Reprinted with permission from: (a) Ref. [42] © 2008 NPG; (b) Ref. [53] © 2018 AIP.

4.3.2 FOCUSED ION-BEAM (FIB) LITHOGRAPHY

Sometimes, EBL will not work well under circumstance when substrate materials are sensitive to solvents or multiple-layer structures require precise alignment. As an alternative maskless nano-fabrication technique, FIB lithography has become a popular tool for nanostructuring high-quality and high-precision metamaterials.

Focused ion-beam lithography techniques are analogous to EBL, but replacing the source of electrons with much heavier ion masses. These heavier high-energy ionized atoms such as Ga^+ are focused onto the sample for physically removing atoms from the substrate surface with very large energy transfer at a rate several orders of magnitude higher than that of electrons. The feature size of patterns is mainly determined by the beam size and the interactions of the beam with the substrate material. It has been shown FIB can achieve sub-5-nm resolution on a thin SiC membrane [44]. Additionally, different from EBL process, FIB is a rapid prototyping process that does not involve any postprocessing, such as resist development, or lift-off steps, leading to a relatively short fabrication time. The whole process is as simple as depositing uniform thin films onto a substrate and etching patterns using FIB.

The FIB technique has been widely used to fabricate plasmonic metamaterials and all-dielectric metamaterials where relatively high aspect-ratio of holes or gaps are needed. In addition, since the selectivity of FIB process is not as high as that of RIE, it is normally used to define structures consisting of several different materials in one-step milling process [42, 45, 46]. Figure 4.4a shows an SEM image of a fabricated 3D negative-index prism consisting of fishnet metamaterials with alternative layers of 30-nm silver and 50-nm magnesium fluoride (MgF_2) [42]. The patterned structure shows a feature size of 265-nm and aspect ratio of over 3.

Using such a multilayer metal-dielectric stack, Valentine et al. successfully demonstrated negative refraction at a wavelength of 1763-nm.

In addition to creating multi-layer stacks consisting of various materials, the FIB technique was recently shown capable of shaping free-standing thin films for construction of reconfigurable metamaterials [47–49, 52, 53]. Through carving out of a free-standing gold-SiN$_x$-gold membrane, Ou et al. demonstrated a reconfigurable photonic metamaterial with transmission modulation up to 50%, due to the heat induced mechanical bulking from the thermal expansion mismatch between gold and silicon nitride [47]. Rather than milling materials from substrates, FIB can also be used to spatially control local stress of membranes to precisely deform structures via creating vacancies and/or implanting gallium ions into the thin film surface. Using this method, Li and his colleagues demonstrated various nano-kirigami metamaterials with cutting and folding free-standing gold thin films, such as standing-up ring resonator arrays [51], 3D vertical helix arrays [52], and diffractive gratings [53]. Figure 4.4b shows the SEM image of a circular grating, which includes left-handed and right-handed pinwheel structures, as shown in the inset.

Similar to the EBL technique, FIB is also a serial process, which leads to a relatively long process time for a square millimeter-area patterning. However, as opposed to EBL or other conventional lithographic techniques, FIB milling is a subtractive process, in which the materials after fabrication is not reversible. If the focused ion beam is not very stable across the etching area during writing, the patterns can be easily ruined, thus leading to a useless sample.

4.3.3   INTERFERENCE LITHOGRAPHY

As mentioned above, EBL and FIB nanofabrication techniques are normally suitable for proof-of-principle studies due to their serial nature, which hinders their applications for large-area patterning and high-volume production with reasonably low cost and time. Fortunately, interference lithography (IL), evolved from conventional optical lithography, can produce periodic or quasi-periodic patterns over large areas, and high volumes. Yet, the multiple exposures and multiple-beam interference, and mix-and-match synthesis with other lithographic techniques could significantly expand patterning flexibility for more complicate patterns [54, 55].

The basic principle of IL is to directly expose resist with a standing wave pattern, which is constructed by the interference of at least two coherent light beams. For a simple two-beam IL, the periodicity of the patterns is determined by the light wavelength, refractive index of the medium, i.e., air or immersion liquid, and angle between the two coherent waves. When combined with an immersion technique, the resolution of IL can approach a scale of 20-nm. With decades of development, IL has found numerous applications in fields other than the semiconductor industry, such as biomedical scaffolds, optical trapping, photonic crystals, and metamaterials.

Given the nature of IL producing periodic or quasi-periodic patterns, it is an excellent candidate for large-area metamaterial production. Through controlling the incidence angles,

(a) 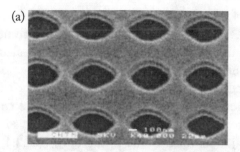 (b)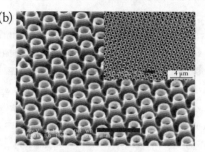

Figure 4.5: (a) SEM image of a fishnet metamaterial with negative index fabricated by the IL process. The pitch of the sample is 787-nm in both directions. The diameter of the elliptical hole in horizontal direction is 540-nm and the diameter in vertical direction is 380-nm. Scale bar is 100-nm. (b) SEM image of IL-fabricated U-shaped arrays with oblique view. Scale bar is 1 μm. The inset is the normal view of the patterns. Reprinted with permission from: (a) Ref. [57] © 2006 OSA; (b) Ref. [61] © 2008 OSA.

intensity, polarization, and relative phases of the laser beams, IL is able to produce a variety of shapes, such as 1D array of grating structure [56], 2D array of holes [57–59], dots, [56, 60], and cut-wire pairs [56], etc. IL was first employed to fabricated arrayed fishnet metamaterials with elliptical holes, as shown in Fig. 4.5a, which exhibit negative refractive index in the near-infrared wavelength [57]. The fishnet metamaterial consists of metal-dielectric-metal composites with a feature size around 380 nm along the minor axis of the elliptical holes.

It is attractive to pattern metamaterials using IL technique because of its easy implementation, low cost and high throughput. However, the interference of beams is inherently restricted to locally periodic patterns because of the nature of the periodic electromagnetic waves. Further, it is also challenging to carve out unit-cell structure with complex shapes other than circles, squares, stripes, etc., using conventional two-beam IL technique. Yang et al. introduced a single-beam holographic lithography to generate diverse structures, including cylindrical nanoshells, U-shaped resonator arrays [61], a symmetrically top-cut hexagonal prism is used to split a collimated beam into seven beams by the top and six side-surfaces and then automatically reassemble the seven beams together under the bottom face of the prism to form the interference patterns. Figure 4.5b shows the fabricated U-shaped metamaterial arrays using the holographic lithography technique.

4.3.4   MULTI-PHOTON POLYMERIZATION LITHOGRAPHY

It is well known that a 2D sub-wavelength metallic metamaterial (normal to the wavevector direction) would only exhibit electric response with tailorable electric polarizability, yet 3D metamaterials could provide further freedoms to manipulate the electromagnetic waves, enabling, for example, polarization control [62], Huygens' metasurface [63], hyperbolic metamaterial for superlensing [46], invisible cloaking [64, 65], etc. However, most of the aforementioned fabrication techniques majorly aim to shape 2D planar metamaterials. Although some of them, such as the conventional and electron-beam lithography methods have been revealed the capability to build up 3D metamaterials via stacking multiple layers [40, 66, 67], they are not plausible for fabricating intricate structures, like spirals.

There have been several fabrication techniques developed to fabricate complex 3D microstructures with potential for mass production and mesoscopic patterning, such as gray-scale lithography [68], stereolithography [69], and multiphoton lithography (MPL), also known as direct laser writing (DWL) technique [70]. Among these techniques, MPL technique is able to provide sub-100-nm resolution and relatively high throughput. Besides, it is also suitable for a variety of patternable materials, including photopolymers, photoresist, and mixed materials of nanoparticles and metal ions. The MPL technique involves a nonlinear process with simultaneously absorbing two or more photons to induce electron transitions in the polymer at the focal point of a tightly focused laser beam, consequently activating the polymerization process. Taking the two-photon polymerization as an example, the rate of the two-photo absorption is proportional to the square of the laser intensity. As a result, the solidification of the polymer, induced by a Gaussian laser spot, is significantly confined in a submicron region, which is beyond the diffraction limit of the working wavelength. In addition, different from stereolithography where a sequential layer-by-layer patterning process has to be implemented, MPL can achieve polymerization at any desired location via computer-controlled scanning of the focus using piezoelectric actuators. Such a technique allows fabrication of freeform 3D structures with lateral resolution as small as 10-nm [71].

Since polymer structures, or voids, can be precisely defined by the beam focus, the DWL technique exhibits the capability to fabricate complicate 3D structure with feature size changing from micrometers to tens of nanometers [65, 72, 73]. Then, through varying the volume filling fraction of the polymer, it is easy to control the local effective refractive index, with values ranging between the refractive index of the polymer and free space. Based on this technique, a 3D near-infrared carpet cloak made of negative photoresist IP-L was experimentally demonstrated with spatially varied spacing between piles of pillars. The invisible cloaking was achieved by coating a 150-nm gold film on the top valley region, mimicking a hidden object. The incidence and reflectance measurements were implemented from the bottom. The optical characterization showed that the visibility of the valley is strongly suppressed in the range from 1.5–2.6 μm [65].

Although most demonstrations of DWL were achieved directly using polymers [74–77], recent studies also showed the feasibility for metallic metamaterials. In 2006, a large-size 3D periodic metallic micro/nanostructure using an electroless plating technique over the fabricated polymer matrix was experimentally demonstrated [78]. To improve the lithography efficiency, a microlens array with periodicity of 300 μm was adopted to pattern over 700 arrayed structures in parallel. Later, a similar method using a chemical vapor deposition (CVD) technique was demonstrated to fabricate a high-quality 3D magnetic metamaterial operating at near-IR wavelength [79]. Figure 4.6a shows an oblique cross-section view of the fabricated structures, which consists of 1D array dome-shape structure with a 50-nm silver thin film. The thin dome shell is supported by an SU-8 polymeric template fabricated by the DWL technique. The dome-shape resonator structure is analogous to the conventional split-ring resonator. To deposit the silver, the template was heated up to 160°C, which is well above the sublime temperature of the metal-organic precursor (COD)(hfac)Ag(I). The vapor deposition allows the metal to be conformally coated around the structure and exhibiting good d.c. conductivity. Given the magnetic response of the structure, additional silver wires along the polarization direction provides a strong electric response. Through vacuum electron-beam evaporation of silver, 3D crossbar structures reveal negative refractive index despite the bianisotropic nature of the metamaterial structure [80]. This approach enables rapid prototyping of high-quality 3D photonic metamaterials.

The electromagnetic response of metallic metamaterials are mainly established by the geometry of patterned metal. However, the selectivity of electroless plating and vapor deposition is relatively poor. Consequently, a synthesis process could take place on all structures, including the substrate [82], resulting in deteriorated electromagnetic response. Instead of using negative-tone photoresist, a positive-tone photoresist AZ9260 has been directly used to fabricated a 3D spiral structure directly [62]. The metallic structure is formed by electroplating gold via the narrow pores on an ITO film. The single-helix structure with only one helix winding can be operated as a broadband circular polarizer. However, due to the lack of the rotational symmetry for a single-helix structure, the circular polarization conversion between left-handed and right-handed polarizations cannot be fully eliminated. Then, to reduce the polarization conversion, a tapered helix structure fabricated by the same method was demonstrated to increase the extinction ratio [83]. To further fully eliminate the circular polarization conversion, Kaschke et al. introduced a multiple-helix metamaterial with three helices intertwining within a unit cell using the stimulated-emission depletion-(STED)–inspired direct laser writing [81]. Similarly, a polymer template with a polymer shell surrounding the desired helical structure was written on a conductive ITO thin film as shown in Fig. 4.6b. Figure 4.6c shows a fabricated triple-helix metamaterials with a pitch size of 2 μm and wire radius of 270 nm. The measured Jones transmission and reflection matrices indicate a broadband circular dichroism of the triple-helix metamaterial with only a small circular polarization conversion.

In addition to polarization conversion and invisibility cloaking, MPL was also applied to demonstrate 3D perfect absorbers in the mid-IR wavelength, as shown in Fig. 3.12l [84]. Distinct from conventional planar metamaterial perfect absorbers, such a design with two standing U-shaped rings supports polarization independent magnetic dipoles on a ground plane. According to temporal coupled-mode theory, the single magnetic dipole on a image plane is also capable to obtain perfect absorption. The experiment results show the fabricated samples achieved 90% absorption around 6 μm. Multiphoton fabrication techniques have shown the ability to create arbitrarily complex 3D structures with sub-micrometer feature sizes. Most metamaterials work is planar, i.e., metasurfaces, owing to easier fabrication and promise for compact optical systems.

(a)

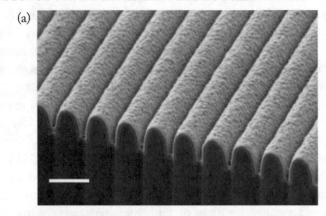

(b)  (c)

Figure 4.6: (a) Oblique view of a magnetic metamaterial structure that has been cut by a focus-ion beam. The dark region is the SU-8 resist and the grey region is the coated silver. (b) and (c) SEM images of the polymer cast and the final gold structures formed by electroplating, respectively. Reprinted with permission from: (a) Ref. [79] © 2008 NPG; (b); and (c) Ref. [81] © 2015 OSA.

However, recent progress on meta-lenses using two or more layers to obtain addition degrees of freedom to manipulate the phase, amplitude and polarization, suggests the promise of 3D configurations for multifunctional metamaterials. Moreover, with the rapid progress of 3D additive manufacturing, 3D structures might be as easy to fabricate as 2D.

4.3.5   SELF-ASSEMBLY LITHOGRAPHY

As discussed in previous sections, EBL and FIB are mature techniques for sub-micrometer fabrication, however, it is laborious to apply them for large-scale metamaterial production due to their series processing nature. Additionally, optical metamaterials have shown intriguing properties for a variety of important applications in visible range, including superresolution imaging, optical cloaking, lasing, and sensing. However, it is highly demanding to fabricate optical metamaterials with feature size of tens of nanometers by conventional top-down lithography approaches. Recently, a novel bottom-up self-assembly technique, mediated by weak noncovalent bonds, such as electrostatic interactions, hydrophobic interactions, van der Waals interactions, and water-mediated hydrogen bonds, has become a powerful tool for fabricating novel 2D and 3D metamaterials [85–93]. Of the various self-assembly methods, block-copolymer (BCP) self-assembly and colloidal self-assembly are considered to be most promising tools for low-cost production of optical metamaterials over large-areas.

BCP assembly is a supramolecular assembly technique, primarily governed by non-covalent interactions underlying the thermodynamics. In the assembly process, the molecules spontaneously aggregate together toward an equilibrium condition, thus yielding controllable morphologies normally with a high degree of structural symmetry, including spheres, cylinders, and gyroids. Gyroid morphologies, are particularly interesting since their chiral axes can lead to strong chirality. An isoprene-*block*-styrene-*block*-ethylene oxide (ISO) BCP was used to obtain 3D chiral metamaterials by self-assembly on a conductive substrate [86]. The ISO BCP is able to form two chemically distinct and interpenetrating gyroid networks with opposite chirality in a matrix. Through selective UV and chemical etching of polyisoprene, followed by electroplating gold into the voids. After removing the surrounding polymer template, a single gyroid network replicating the original structure of polyisoprene block, as shown in Fig. 4.7a. Fundamentally limited by the thermodynamics, the self-assembled morphologies are constrained on the lattice configurations and symmetry. However, through introduction of external excitations or selectively modifying linking chemicals, it is possible to overcome the thermodynamic restriction, leading to tunable metamaterials with tailored symmetries and EM response. With transferring BCP self-assembled metal particles onto a thermal shrinkage film, the subsequent pattern shrinkage allows spatially control lattice spacing and symmetry precisely, resulting in significant refractive index increase up to 5.1 at visible wavelength [91]. Additionally, it has been shown that the self-assembled plasmonic nanorod dimers can be controlled with a longitudinal offset using a feedback mechanism, in which assemblies can be disassembled through inducing photothermal effect in the unwanted dimers by selective light excitation. As a result, a reconfigured and homogenized aqueous metamaterial with asymmetrically assembly plasmonic dimers can be obtained [87].

Colloidal self-assembly is a synthetic method for preparing monodisperse polymer colloids in 2D and 3D configurations using microspheres or nanospheres. Recently, colloidal self-assembly has attracted much interest due to their low cost, high efficiency for large-area patterning, and their potentials in many important applications, such as imaging [95, 96], sensing [97], and spectroscopy [98]. 2D colloidal crystals are usually assembled on planar substrates or at an air/water interface to form a close-packed configurations as a result of the reduction of free energy to approach local equilibrium. Recent developments in colloidal crystals have

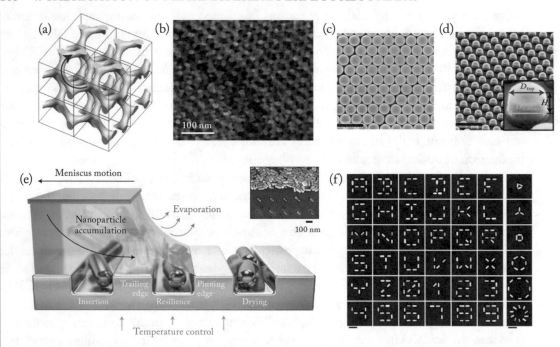

Figure 4.7: (a) A computer-simulated view of a single gyroid with a clockwise handedness as indicated by the arrow. (b) SEM image of a metal gyroid after filling a porous polymer network. (c) SEM image of close-packed polystyrene spheres with diameter of 820-nm. (d) A tilted view of the fabricated all-dielectric metasurface consisting of an array of silicon cylinders (blue, false color). The inset shows an SEM image of a single Si resonator with $D_{top} = 480$-nm, $D_{bottom} = 554$-nm, and height $H = 335$-nm. The scale bar is 2 μm. (e) Schematic of the template-assisted capillary assembly of nanoparticles on a tomographic template. The inset shows the assembled nanorods. (f) SEM images of some examples of patterned gold nanorods by template-assisted capillary assembly. The left 6 columns include the 26 characters of the Latin alphabet and the Arabic numbers. The right column shows the some hierarchical ring clusters with various number of nanorods. Reprinted with permission from: (a) and (b) Ref. [86] © 2012 Wiley-VCH; (c) and (d) Ref. [89] © 2015 ACS; (e) and (f) Ref. [102] © 2016 NPG.

enabled self-assembly of plasmonic nanoparticles with controlled patterns, thus leading to significant electromagnetic responses [99–101]. Driven by the capillary force, hierarchical plasmonic structures with various numbers of particles inside the clusters have been demonstrated to exhibit strong electric, magnetic, and Fano-like resonances. The assembled clusters consists of gold shell coated silica nanoparticles with controlled spacing between particles as small as 2 nm, which is beyond the ability of conventional lithography. However, the randomly distributed size and position of self-assembled clusters hinders them for large-area spectroscopy and practical applications. To achieve precise control of colloidal self-assembly with well-defined structures and spatial order, approaches including templates [101, 102], external fields [103, 104], and directing agents [105, 106] have been investigated. Template-assisted capillary assembly is a technique that enables precise control of position, orientation and interparticle spacing at the nanometer level for self-assembled nanoparticles. Figure 4.7e shows a schematic of the template-assisted capillary assembly in which a controlled volume of colloidal solution is confined between a patterned, low-wetting template, while the top plate is set in a relative sliding motion [102]. Through control of the rate of evaporation of the meniscus front, the particles are selectively trapped in the topographic features. The inset of Fig. 4.7f shows some examples of patterns by template-assisted capillary assembly, including 26 characters, Arabic numbers, and some clusters.

In addition to directly obtaining plasmonic patterns, colloidal self-assembly technique can be also used for colloidal lithography, also known as sphere lithography, where the orderly packed colloidal crystals are used as a hard mask for patterning arrayed structures. Generally, polystyrene (PS) spheres are placed in a Langmuir–Blodgett container to form a close-packed hexagonal (hcp) monolayer at the interface between air and water [107]. Then, the self-assembled layer is transferred to a target substrate for subsequent etching or deposition process, as shown in Fig. 4.7c. The colloidal lithographic technique has been well demonstrated for patterning plasmonic metasurface in the infrared and visible wavelength [108–112]. Recently, this technique was also shown capable for fabricating all-dielectric metasurfaces. Figure 4.7d shows the SEM image of the transferred hcp PS spheres with diameter of 820-nm [89]. After reactive ion etching on the spheres and the SOI substrate, the arrayed cylindrical disks exhibits near-perfect reflection due the excited magnetic dipole resonance leading to a zero phase change after reflection, which is drastically different from a electric dipole resonance induced reflection with a phase change close to 180°, like a metallic mirror. Not only patterned on a rigid substrate, all-dielectric metasurface also can be directly fabricated on a flexible substrate [113].

The self-assembly approach has been shown to be facile, low cost, and efficient to fabricate 2D and 3D metamaterials. These distinct advantages over conventional top-down methods make the self-assembly technique favored for practical applications involving large-area optical metamaterials. Recently, self-assembly nanosphere lithography technique was adopted to fabricate an infrared all-dielectric metasurface with dimensions of 3.5×3.5 mm^2 for edge detection [114]. Although great progress in the self-assembly technique has been made for metamaterials, it is still far from a well-developed method, since it is still challenging to pattern metamaterials without regular shapes or periodicity, such as flat metalens [115], and holographic metasurface [116], in which the metamaterial unit cells are spatially varied.

4.4 CONCLUSION

The development of the metamaterials field has undoubtedly benefited from the progress of fabrication techniques. It is important to note that the fabricational techniques used for demonstra-

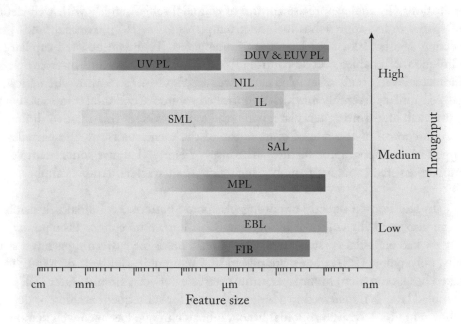

Figure 4.8: Summarized feature size and throughput of major lithography techniques used for metamaterial fabrications. UV PL: ultraviolet photolithography; DUV: deep ultraviolet; EUV: extreme ultraviolet; NIL: nanoimprinting lithography; IL: interference lithography; SML: shadow-mask lithography; SAL: self-assembly lithography; MPL: multi-photon polymerization lithography; EBL: electron-beam lithography; FIB: focused ion beam.

tion of metamaterials is not limited to what was shown here. Some other alternative techniques, such as plasmonic lithography [117, 118], off-normal deposition [119], laser ablation [120], etc., are also well known for metamaterial patterning. Currently, studies of metamaterials have been gradually shifting from fundamental to more applied. Recent demonstrations of wavefront shaping and polarization control using metamaterials have shown remarkable potential to manipulate electromagnetic waves beyond conventional optical components. It is clear that fabricational techniques with large area, high resolution, low cost, high throughput, good reproducibility, and reliability are crucial for metamaterial success in industry. Figure 4.8 summarizes the feature sizes achievable for various lithographic techniques, as well as provides a relative comparison of their throughput for metamaterial patterning [121–123]. It is expected that continued development of fabricational techniques will enable metamaterial applications to disrupt the optics industry in the near future.

4.5 REFERENCES

[1] Julius Adams Stratton. *Electromagnetic Theory*. John Wiley & Sons, Inc., October 2015. DOI: 10.1002/9781119134640 93

[2] Lucas H. Gabrielli, Jaime Cardenas, Carl B. Poitras, and Michal Lipson. Silicon nanostructure cloak operating at optical frequencies. *Nature Photonics*, 3(8):461–463, August 2009. DOI: 10.1038/nphoton.2009.117 93

[3] Shuang Zhang, Yong-Shik Park, Jensen Li, Xinchao Lu, Weili Zhang, and Xiang Zhang. Negative refractive index in chiral metamaterials. *Physical Review Letters*, 102:023901, January 2009. DOI: 10.1103/physrevlett.102.023901 93

[4] Shumin Xiao, Uday K. Chettiar, Alexander V. Kildishev, Vladimir P. Drachev, and Vladimir M. Shalaev. Yellow-light negative-index metamaterials. *Optics Letters*, 34(22):3478–3480, November 2009. DOI: 0.1364/ol.34.003478 93

[5] Hu Tao, Nathan I. Landy, Christopher M. Bingham, Xin Zhang, Richard D. Averitt, and Willie J. Padilla. A metamaterial absorber for the terahertz regime: Design, fabrication and characterization. *Optics Express*, 16(10):7181, May 2008. DOI: 10.1364/oe.16.007181 93

[6] Hu Tao, C. M. Bingham, A. C. Strikwerda, D. Pilon, D. Shrekenhamer, N. I. Landy, K. Fan, X. Zhang, W. J. Padilla, and R. D. Averitt. Highly flexible wide angle of incidence terahertz metamaterial absorber: Design, fabrication, and characterization. *Physical Review B*, 78(24):2–5, December 2008. DOI: 10.1103/physrevb.78.241103 93

[7] Xianliang Liu, Tatiana Starr, Anthony F. Starr, and Willie J. Padilla. Infrared spatial and frequency selective metamaterial with near-unity absorbance. *Physical Review Letters*, 104(20):207403, May 2010. DOI: 10.1103/physrevlett.104.207403 93

[8] Jiaming Hao, Jing Wang, Xianliang Liu, Willie J. Padilla, Lei Zhou, and Min Qiu. High performance optical absorber based on a plasmonic metamaterial. *Applied Physics Letters*, 96(25):251104, June 2010. DOI: 10.1063/1.3442904 93

[9] Manuel Decker, Isabelle Staude, Matthias Falkner, Jason Dominguez, Dragomir N. Neshev, Igal Brener, Thomas Pertsch, and Yuri S. Kivshar. High-efficiency dielectric Huygens' surfaces. *Advanced Optical Materials*, 3(6):813–820, June 2015. DOI: 10.1002/adom.201400584 93

[10] Kebin Fan, Jingdi Zhang, Xinyu Liu, Gu-Feng Zhang, Richard D. Averitt, and Willie J. Padilla. Phototunable dielectric Huygens' metasurfaces. *Advanced Materials*, 30(22):1800278, May 2018. DOI: 10.1002/adma.201800278 93

[11] Marc J. Madou. *Fundamentals of Microfabrication and Nanotechnology*, 3rd ed., CRC Press, 2018. DOI: 10.1201/9781315274164 94

[12] Uzodinma Okoroanyanwu. Invention of lithography and photolithography. In *Chemistry and Lithography*, pages 9–27, SPIE, 2020. DOI: 10.1117/3.821384.ch2 94

[13] Hu Tao, Jason J. Amsden, Andrew C. Strikwerda, Kebin Fan, David L. Kaplan, Xin Zhang, Richard D. Averitt, and Fiorenzo G. Omenetto. Metamaterial silk composites at terahertz frequencies. *Advanced Materials*, 22(32):3527–3531, August 2010. DOI: 10.1002/adma.201000412 95, 96

[14] Serap Aksu, Min Huang, Alp Artar, Ahmet A. Yanik, Selvapraba Selvarasah, Mehmet R. Dokmeci, and Hatice Altug. Flexible plasmonics on unconventional and non-planar substrates. *Advanced Materials*, 23(38):4422–4430, October 2011. DOI: 10.1002/adma.201102430 95, 96

[15] Tao Deng, Mengwei Li, Jian Chen, Yifan Wang, and Zewen Liu. Controllable fabrication of pyramidal silicon nanopore arrays and nanoslits for nanostencil lithography. *The Journal of Physical Chemistry C*, 118(31):18110–18115, August 2014. DOI: 10.1021/jp503203b 95

[16] Mengkun Liu, Harold Y. Hwang, Hu Tao, Andrew C. Strikwerda, Kebin Fan, George R. Keiser, Aaron J. Sternbach, Kevin G. West, Salinporn Kittiwatanakul, Jiwei Lu, Stuart A. Wolf, Fiorenzo G. Omenetto, Xin Zhang, Keith A. Nelson, and Richard D. Averitt. Terahertz-field-induced insulator-to-metal transition in vanadium dioxide metamaterial. *Nature*, 487(7407):345–8, July 2012. DOI: 10.1038/nature11231 95, 96

[17] Wenzhong Bao, Gang Liu, Zeng Zhao, Hang Zhang, Dong Yan, Aparna Deshpande, Brian LeRoy, and Chun Ning Lau. Lithography-free fabrication of high quality substrate-supported and freestanding graphene devices. *Nano Research*, 3(2):98–102, February 2010. DOI: 10.1007/s12274-010-1013-5 95

[18] M. Gurvitch, S. Luryi, A. Polyakov, A. Shabalov, M. Dudley, G. Wang, S. Ge, and V. Yakovlev. VO2 films with strong semiconductor to metal phase transition prepared by the precursor oxidation process. *Journal of Applied Physics*, 102(3):033504, August 2007. DOI: 10.1063/1.2764245 95

[19] M. Tayyab Nouman, Ji Hyun Hwang, Mohd. Faiyaz, Kye-Jeong Lee, Do-Young Noh, and Jae-Hyung Jang. Vanadium dioxide based frequency tunable metasurface filters for realizing reconfigurable terahertz optical phase and polarization control. *Optics Express*, 26(10):12922, May 2018. DOI: 10.1364/oe.26.012922 95

[20] Andrew C. Strikwerda, Kebin Fan, Hu Tao, Daniel V. Pilon, Xin Zhang, and Richard D. Averitt. Comparison of birefringent electric split-ring resonator and meanderline structures as quarter-wave plates at terahertz frequencies. *Optics Express*, 17(1):136, December 2008. DOI: 10.1364/oe.17.000136 95

[21] W.-C. Chen, N. I. Landy, K. Kempa, and W. J. Padilla. A subwavelength extraordinary-optical-transmission channel in babinet metamaterials. *Advanced Optical Materials*, 1(3):221–226, March 2013. DOI: 10.1002/adom.201200016 95

[22] Katrin Sidler, Luis G. Villanueva, Oscar Vazquez-Mena, Veronica Savu, and Juergen Brugger. Compliant membranes improve resolution in full-wafer micro/nanostencil lithography. *Nanoscale*, 4(3):773–778, January 2012. DOI: 10.1039/c2nr11609j 96

[23] Jun Cai, Xiaoning Wang, Aobo Li, Stephan W. Anderson, and Xin Zhang. Biologically enabled micro- and nanostencil lithography using diatoms. *Extreme Mechanics Letters*, 4:186–192, September 2015. DOI: 10.1016/j.eml.2015.07.003 96

[24] Stephen Y. Chou, Peter R. Krauss, and Preston J. Renstrom. Imprint of sub-25 nm vias and trenches in polymers. *Applied Physics Letters*, 67(21):3114–3116, November 1995. DOI: 10.1063/1.114851 97

[25] S. Y. Chou, P. R. Krauss, and P. J. Renstrom. Imprint lithography with 25-nanometer resolution. *Science*, 272(5258):85–87, April 1996. DOI: 10.1126/science.272.5258.85 97

[26] L. J. Guo. Nanoimprint lithography: Methods and material requirements. *Advanced Materials*, 19(4):495–513, February 2007. DOI: 10.1002/adma.200600882 97

[27] Nazrin Kooy, Khairudin Mohamed, Lee Pin, and Ooi Guan. A review of roll-to-roll nanoimprint lithography. *Nanoscale Research Letters*, 9(1):320, June 2014. DOI: 10.1186/1556-276x-9-320 97

[28] Yifang Chen, Jiarui Tao, Xingzhong Zhao, Zheng Cui, Alexander S. Schwanecke, and Nikolay I. Zheludev. Nanoimprint lithography for planar chiral photonic meta-materials. *Microelectronic Engineering*, 78-79:612–617, March 2005. DOI: 10.1016/j.mee.2004.12.078 97, 98

[29] Wei Wu, Zhaoning Yu, Shih-Yuan Wang, R. Stanley Williams, Yongmin Liu, Cheng Sun, Xiang Zhang, Evgenia Kim, Y. Ron Shen, and Nicholas X. Fang. Midinfrared metamaterials fabricated by nanoimprint lithography. *Applied Physics Letters*, 90(6):063107, February 2007. DOI: 10.1063/1.2450651 97

[30] W. Wu, E. Kim, E. Ponizovskaya, Y. Liu, Z. Yu, N. Fang, Y. R. Shen, A. M. Bratkovsky, W. Tong, C. Sun, X. Zhang, S.-Y. Wang, and R. S. Williams. Optical metamaterials at near and mid-IR range fabricated by nanoimprint lithography. *Applied Physics A*, 87(2):143–150, March 2007. DOI: 10.1007/s00339-006-3834-3 97

[31] I. Bergmair, B. Dastmalchi, M. Bergmair, A. Saeed, W. Hilber, G. Hesser, C. Helgert, E. Pshenay-Severin, T. Pertsch, E. B. Kley, U. Hübner, N. H. Shen, R. Penciu, M. Kafesaki, C. M. Soukoulis, K. Hingerl, M. Muehlberger, and R. Schoeftner. Single and multilayer metamaterials fabricated by nanoimprint lithography. *Nanotechnology*, 22(32):325301, August 2011. DOI: 10.1088/0957-4484/22/32/325301 97

[32] Jong G. Ok, Hong Seok Youn, Moon Kyu Kwak, Kyu-Tae Lee, Young Jae Shin, L. Jay Guo, Anton Greenwald, and Yisi Liu. Continuous and scalable fabrication of flexible metamaterial films via roll-to-roll nanoimprint process for broadband plasmonic infrared filters. *Applied Physics Letters*, 101(22):223102, November 2012. DOI: 10.1063/1.4767995 97, 98

[33] Stephen Y. Chou and Peter R. Krauss. Imprint lithography with sub-10 nm feature size and high throughput. *Microelectronic Engineering*, 35(1–4):237–240, February 1997. DOI: 10.1016/s0167-9317(96)00097-4 97

[34] Wei Wu, William M. Tong, Jonathan Bartman, Yufeng Chen, Robert Walmsley, Zhaoning Yu, Qiangfei Xia, Inkyu Park, Carl Picciotto, Jun Gao, Shih-Yuan Wang, Deborah Morecroft, Joel Yang, Karl K. Berggren, and R. Stanley Williams. Sub-10 nm nanoimprint lithography by wafer bowing. *Nano Letters*, 8(11):3865–3869, November 2008. DOI: 10.1021/nl802295n 97

[35] Tomi Haatainen, Päivi MaJander, Tommi Riekkinen, and Jouni Ahopelto. Nickel stamp fabrication using step and stamp imprint lithography. *Microelectronic Engineering*, 83(4–9):948–950, April 2006. DOI: 10.1016/j.mee.2006.01.038 97

[36] Suho Ahn, Joowon Cha, Ho Myung, Seok-min Kim, and Shinill Kang. Continuous ultraviolet roll nanoimprinting process for replicating large-scale nano- and micropatterns. *Applied Physics Letters*, 89(21):213101, November 2006. DOI: 10.1063/1.2392960 99

[37] Se Hyun Ahn and L. Jay Guo. High-speed roll-to-roll nanoimprint lithography on flexible plastic substrates. *Advanced Materials*, 20(11):2044–2049, June 2008. DOI: 10.1002/adma.200702650 99

[38] Stefan Linden, Christian Enkrich, Martin Wegener, Jiangfeng Zhou, Thomas Koschny, and Costas M. Soukoulis. Magnetic response of metamaterials at 100 terahertz. *Science*, 306(5700):1351–3, November 2004. DOI: 10.1126/science.1105371 100

[39] C. Enkrich, M. Wegener, S. Linden, S. Burger, L. Zschiedrich, F. Schmidt, J. Zhou, Th. Koschny, and C. Soukoulis. Magnetic metamaterials at telecommunication and visible frequencies. *Physical Review Letters*, 95(20):203901, November 2005. DOI: 10.1103/physrevlett.95.203901 100

[40] Na Liu, Hongcang Guo, Liwei Fu, Stefan Kaiser, Heinz Schweizer, and Harald Giessen. Three-dimensional photonic metamaterials at optical frequencies. *Nature Materials*, 7(1):31–37, January 2008. DOI: 10.1038/nmat2072 100, 104

[41] Stéphane Larouche, Yu-Ju Tsai, Talmage Tyler, Nan M. Jokerst, and David R. Smith. Infrared metamaterial phase holograms. *Nature Materials*, 11(5):450–454, May 2012. DOI: 10.1038/nmat3278 100

[42] Jason Valentine, Shuang Zhang, Thomas Zentgraf, Erick Ulin-Avila, Dentcho A. Genov, Guy Bartal, and Xiang Zhang. Three-dimensional optical metamaterial with a negative refractive index. *Nature*, 455(7211):376–9, September 2008. DOI: 10.1038/nature07247 101

[43] Zhiguang Liu, Huifeng Du, Zhi-Yuan Li, Nicholas X. Fang, and Jiafang Li. Invited article: Nano-kirigami metasurfaces by focused-ion-beam induced close-loop transformation. *APL Photonics*, 3(10):100803, October 2018. DOI: 10.1063/1.5043065

[44] J. Gierak, A. Madouri, A.L. Biance, E. Bourhis, G. Patriarche, C. Ulysse, D. Lucot, X. Lafosse, L. Auvray, L. Bruchhaus, and R. Jede. Sub-5 nm FIB direct patterning of nanodevices. *Microelectronic Engineering*, 84(5–8):779–783, May 2007. DOI: 10.1016/j.mee.2007.01.059 101

[45] Nicholas Fang, Hyesog Lee, Cheng Sun, and Xiang Zhang. Sub-diffraction-limited optical imaging with a silver superlens. *Science*, 308(5721):534–7, April 2005. DOI: 10.1126/science.1108759 101

[46] Zhaowei Liu, Hyesog Lee, Yi Xiong, Cheng Sun, and Xiang Zhang. Far-field optical hyperlens magnifying sub-diffraction-limited objects. *Science*, 315(5819):1686, March 2007. DOI: 10.1126/science.1137368 101, 104

[47] J. Y. Ou, E. Plum, L. Jiang, and N. I. Zheludev. Reconfigurable photonic metamaterials. *Nano Letters*, 11(5):2142–4, May 2011. DOI: 10.1021/nl200791r 102

[48] Jun-Yu Ou, Eric Plum, Jianfa Zhang, and Nikolay I. Zheludev. An electromechanically reconfigurable plasmonic metamaterial operating in the near-infrared. *Nature Nanotechnology*, 8(4):252–255, April 2013. DOI: 10.1038/nnano.2013.25 102

[49] João Valente, Jun-Yu Ou, Eric Plum, Ian J. Youngs, and Nikolay I. Zheludev. A magneto-electro-optical effect in a plasmonic nanowire material. *Nature Communications*, 6(1):7021, November 2015. DOI: 10.1038/ncomms8021 102

[50] Zhiguang Liu, Huifeng Du, Jiafang Li, Ling Lu, Zhi-Yuan Li, and Nicholas X. Fang. Nano-kirigami with giant optical chirality. *Science Advances*, 4(7):eaat4436, July 2018. DOI: 10.1126/sciadv.aat4436

[51] Ajuan Cui, Zhe Liu, Jiafang Li, Tiehan H Shen, Xiaoxiang Xia, Zhiyuan Li, Zhijie Gong, Hongqiang Li, Benli Wang, Junjie Li, Haifang Yang, Wuxia Li, and Changzhi Gu. Directly patterned substrate-free plasmonic "nanograter" structures with unusual Fano resonances. *Light: Science and Applications*, 4(7):e308–e308, July 2015. DOI: 10.1038/lsa.2015.81 102

[52] Dianjing Liu, Yixuan Tan, Erfan Khoram, and Zongfu Yu. Training deep neural networks for the inverse design of nanophotonic structures. *ACS Photonics*, 5(4):1365–1369, April 2018. DOI: 10.1021/acsphotonics.7b01377 102

[53] Zhaocheng Liu, Dayu Zhu, Sean P. Rodrigues, Kyu-Tae Lee, and Wenshan Cai. Generative model for the inverse design of metasurfaces. *Nano Letters*, 18(10):6570–6576, October 2018. DOI: 10.1021/acs.nanolett.8b03171 101, 102

[54] S. R. J. Brueck. Optical and interferometric lithography - nanotechnology enablers. *Proc. of the IEEE*, 93(10):1704–1721, October 2005. DOI: 10.1109/jproc.2005.853538 102

[55] Joel Henzie, Min Hyung Lee, and Teri W. Odom. Multiscale patterning of plasmonic metamaterials. *Nature Nanotechnology*, 2(9):549–554, September 2007. DOI: 10.1038/nnano.2007.252 102

[56] Nils Feth, Christian Enkrich, Martin Wegener, and Stefan Linden. Large-area magnetic metamaterials via compact interference lithography. *Optics Express*, 15(2):501, January 2007. DOI: 10.1364/oe.15.000501 103

[57] Shuang Zhang, Wenjun Fan, Kevin J. Malloy, Steven R. J. Brueck, Nicolae C. Panoiu, and Richard M. Osgood. Demonstration of metal-dielectric negative-index metamaterials with improved performance at optical frequencies. *Journal of the Optical Society of America B*, 23(3):434, March 2006. DOI: 10.1364/josab.23.000434 103

[58] Zahyun Ku and S. R. J. Brueck. Comparison of negative refractive index materials with circular, elliptical and rectangular holes. *Optics Express*, 15(8):4515, April 2007. DOI: 10.1364/oe.15.004515 103

[59] Y. Zhou, X. Y. Chen, Y. H. Fu, G. Vienne, A. I. Kuznetsov, and B. Luk'yanchuk. Fabrication of large-area 3D optical fishnet metamaterial by laser interference lithography. *Applied Physics Letters*, 103(12):123116, September 2013. DOI: 10.1063/1.4821508 103

[60] L. J. Heyderman, H. H. Solak, C. David, D. Atkinson, R. P. Cowburn, and F. Nolting. Arrays of nanoscale magnetic dots: Fabrication by x-ray interference lithography and characterization. *Applied Physics Letters*, 85(21):4989–4991, November 2004. DOI: 10.1063/1.1821649 103

[61] Yi Yang, Qiuze Li, and Guo Ping Wang. Design and fabrication of diverse metamaterial structures by holographic lithography. *Optics Express*, 16(15):11275, July 2008. DOI: 10.1364/oe.16.011275 103

[62] Justyna K. Gansel, Michael Thiel, Michael S. Rill, Manuel Decker, Klaus Bade, Volker Saile, Georg von Freymann, Stefan Linden, and Martin Wegener. Gold helix photonic metamaterial as broadband circular polarizer. *Science*, 325(5947):1513–5, September 2009. DOI: 10.1126/science.1177031 103, 105

[63] Carl Pfeiffer and Anthony Grbic. Metamaterial huygens' surfaces: Tailoring wave fronts with reflectionless sheets. *Physical Review Letters*, 110:197401, May 2013. DOI: 10.1103/physrevlett.110.197401 103

[64] Jason Valentine, Jensen Li, Thomas Zentgraf, Guy Bartal, and Xiang Zhang. An optical cloak made of dielectrics. *Nature Materials*, 8(7):568–571, July 2009. DOI: 10.1038/nmat2461 104

[65] Tolga Ergin, Nicolas Stenger, Patrice Brenner, John B. Pendry, and Martin Wegener. Three-dimensional invisibility cloak at optical wavelengths. *Science*, 328(5976):337–339, April 2010. DOI: 10.1126/science.1186351 104

[66] K. B. Fan, A. C. Strikwerda, H. Tao, X. Zhang, and R. D. Averitt. Stand-up magnetic metamaterials at terahertz frequencies. *Optics Express*, 19(13):12619–12627, 2011. DOI: 10.1364/oe.19.012619 104

[67] Kebin Fan, Andrew C. Strikwerda, Xin Zhang, and Richard D. Averitt. Three-dimensional broadband tunable terahertz metamaterials. *Physical Review B*, 87(16):161104, April 2013. DOI: 10.1103/physrevb.87.161104 104

[68] Hideo Kodama. Automatic method for fabricating a three-dimensional plastic model with photo-hardening polymer. *Review of Scientific Instruments*, 52(11):1770–1773, November 1981. DOI: 10.1063/1.1136492 104

[69] X Zhang, X. N. Jiang, and C. Sun. Micro-stereolithography of polymeric and ceramic microstructures. *Sensors and Actuators A: Physical*, 77(2):149–156, October 1999. DOI: 10.1016/s0924-4247(99)00189-2 104

[70] S. Maruo and J. T. Fourkas. Recent progress in multiphoton microfabrication. *Laser and Photonics Review*, 2(1–2):100–111, April 2008. DOI: 10.1002/lpor.200710039 104

[71] Sihao Wang, Ye Yu, Hailong Liu, Kevin T. P. Lim, Bharathi Madurai Srinivasan, Yong Wei Zhang, and Joel K. W. Yang. Sub-10-nm suspended nano-web formation by direct laser writing. *Nano Futures*, 2(2):025006, May 2018. DOI: 10.1088/2399-1984/aabb94 104

[72] Andrew J. Gross and Katia Bertoldi. Additive manufacturing of nanostructures that are delicate, complex, and smaller than ever. *Small*, 15(33):1902370, June 2019. DOI: 10.1002/smll.201902370 104

[73] Dong Yang, Lipu Liu, Qihuang Gong, and Yan Li. Rapid two-photon polymerization of an arbitrary 3D microstructure with 3D focal field engineering. *Macromolecular Rapid Communications*, 40(8):1900041, April 2019. DOI: 10.1002/marc.201900041 104

[74] Brian H. Cumpston, Sundaravel P. Ananthavel, Stephen Barlow, Daniel L. Dyer, Jeffrey E. Ehrlich, Lael L. Erskine, Ahmed A. Heikal, Stephen M. Kuebler, I. Y.Sandy Lee, Dianne McCord-Maughon, Jinqui Qin, Harald Röckel, Mariacristina Rumi, Xiang Li Wu, Seth R. Marder, and Joseph W. Perry. Two-photon polymerization initiators for three-dimensional optical data storage and microfabrication. *Nature*, 398(6722):51–54, March 1999. DOI: 10.1038/17989 104

[75] S. Kawata, H. B. Sun, T. Tanaka, and K. Takada. Finer features for functional microdevices. *Nature*, 412(6848):697–698, August 2001. DOI: 10.1038/35089130 104

[76] Kenji Takada, Hong Bo Sun, and Satoshi Kawata. Improved spatial resolution and surface roughness in photopolymerization-based laser nanowriting. *Applied Physics Letters*, 86(7):1–3, February 2005. DOI: 10.1063/1.1864249 104

[77] S. Passinger, M. S. M. Saifullah, C. Reinhardt, K. R. V. Subramanian, B. N. Chichkov, and M. E. Welland. Direct 3D patterning of TiO2 using femtosecond laser pulses. *Advanced Materials*, 19(9):1218–1221, May 2007. DOI: 10.1002/adma.200602264 104

[78] Florian Formanek, Nobuyuki Takeyasu, Takuo Tanaka, Kenta Chiyoda, Atsushi Ishikawa, and Satoshi Kawata. Three-dimensional fabrication of metallic nanostructures over large areas by two-photon polymerization. *Optics Express*, 14(2):800, January 2006. DOI: 10.1364/opex.14.000800 104

[79] M. S. Rill, C. Plet, M. Thiel, I. Staude, G. Von Freymann, S. Linden, and M. Wegener. Photonic metamaterials by direct laser writing and silver chemical vapour deposition. *Nature Materials*, 7(7):543–546, 2008. DOI: 10.1038/nmat2197 105, 106

[80] Michael S. Rill, Christine E. Kriegler, Michael Thiel, Georg von Freymann, Stefan Linden, and Martin Wegener. Negative-index bianisotropic photonic metamaterial fabricated by direct laser writing and silver shadow evaporation. *Optics Letters*, 34(1):19, January 2009. DOI: 10.1364/ol.34.000019 105

[81] Johannes Kaschke and Martin Wegener. Gold triple-helix mid-infrared metamaterial by STED-inspired laser lithography. *Optics Letters*, 40(17):3986, September 2015. DOI: 10.1364/ol.40.003986 105, 106

[82] Ioanna Sakellari, Xinghui Yin, Maxim L. Nesterov, Konstantina Terzaki, Angelos Xomalis, and Maria Farsari. 3D chiral plasmonic metamaterials fabricated by direct laser writing: The twisted omega particle. *Advanced Optical Materials*, 5(16):1700200, August 2017. DOI: 10.1002/adom.201700200 105

[83] Justyna K. Gansel, Michael Latzel, Andreas Frölich, Johannes Kaschke, Michael Thiel, and Martin Wegener. Tapered gold-helix metamaterials as improved circular polarizers. *Applied Physics Letters*, 100(10):101109, March 2012. DOI: 10.1063/1.3693181 105

[84] Xiang Xiong, Shang-Chi Jiang, Yu-Hui Hu, Ru-Wen Peng, and Mu Wang. Structured Metal Film as a Perfect Absorber. *Advanced Materials*, 25(29):3994–4000, August 2013. DOI: 10.1002/adma.201300223 105

[85] Anton Kuzyk, Robert Schreiber, Zhiyuan Fan, Günther Pardatscher, Eva Maria Roller, Alexander Högele, Friedrich C. Simmel, Alexander O. Govorov, and Tim Liedl. DNA-based self-assembly of chiral plasmonic nanostructures with tailored optical response. *Nature*, 483(7389):311–314, March 2012. DOI: 10.1038/nature10889 107

[86] Silvia Vignolini, Nataliya A. Yufa, Pedro S. Cunha, Stefan Guldin, Ilia Rushkin, Morgan Stefik, Kahyun Hur, Ulrich Wiesner, Jeremy J. Baumberg, and Ullrich Steiner. A 3D optical metamaterial made by self-assembly. *Advanced Materials*, 24(10):OP23–OP27, March 2012. DOI: 10.1002/adma.201103610 107, 108

[87] Sui Yang, Xingjie Ni, Xiaobo Yin, Boubacar Kante, Peng Zhang, Jia Zhu, Yuan Wang, and Xiang Zhang. Feedback-driven self-assembly of symmetry-breaking optical metamaterials in solution. *Nature Nanotechnology*, 9(12):1002–1006, January 2014. DOI: 10.1038/nnano.2014.243 107

[88] Kyuyoung Bae, Gumin Kang, Suehyun K. Cho, Wounjhang Park, Kyoungsik Kim, and Willie J. Padilla. Flexible thin-film black gold membranes with ultrabroadband plasmonic nanofocusing for efficient solar vapour generation. *Nature Communications*, 6:10103, December 2015. DOI: 10.1038/ncomms10103 107

[89] Parikshit Moitra, Brian A. Slovick, Wei Li, Ivan I. Kravchencko, Dayrl P. Briggs, S. Krishnamurthy, and Jason Valentine. Large-scale all-dielectric metamaterial perfect reflectors. *ACS Photonics*, 2(6):692–698, June 2015. DOI: 10.1021/acsphotonics.5b00148 107, 108, 109

[90] Gleb M. Akselrod, Jiani Huang, Thang B. Hoang, Patrick T. Bowen, Logan Su, David R. Smith, and Maiken H. Mikkelsen. Large-area metasurface perfect absorbers from visible to near-infrared. *Advanced Materials*, 27(48):8028–8034, December 2015. DOI: 10.1002/adma.201503281 107

[91] Hyung Ki Kim, Dongju Lee, and Sungjoon Lim. Wideband-switchable metamaterial absorber using injected liquid metal. *Scientific Reports*, 6(1):31823, October 2016. DOI: 10.1038/srep31823 107

[92] Mingkai Liu, Kebin Fan, Willie Padilla, David A. Powell, Xin Zhang, and Ilya V. Shadrivov. Tunable meta-liquid crystals. *Advanced Materials*, 28(8):1553–1558, February 2016. DOI: 10.1002/adma.201504924 107

[93] Jon W. Stewart, Gleb M. Akselrod, David R. Smith, and Maiken H. Mikkelsen. Toward multispectral imaging with colloidal metasurface pixels. *Advanced Materials*, 29(6):1602971, February 2017. DOI: 10.1002/adma.201602971 107

[94] Ju Young Kim, Hyowook Kim, Bong Hoon Kim, Taeyong Chang, Joonwon Lim, Hyeong Min Jin, Jeong Ho Mun, Young Joo Choi, Kyungjae Chung, Jonghwa Shin, Shanhui Fan, and Sang Ouk Kim. Highly tunable refractive index visible-light metasurface from block copolymer self-assembly. *Nature Communications*, 7(1):1–9, September 2016. DOI: 10.1038/ncomms12911

[95] André C. Arsenault, Daniel P. Puzzo, Ian Manners, and Geoffrey A. Ozin. Photonic-crystal full-colour displays. *Nature Photonics*, 1(8):468–472, August 2007. DOI: 10.1038/nphoton.2007.140 107

[96] Shin-Hyun Kim, Seog-Jin Jeon, Woong Chan Jeong, Hyo Sung Park, and Seung-Man Yang. Optofluidic synthesis of electroresponsive photonic Janus balls with isotropic structural colors. *Advanced Materials*, 20(21):NA–NA, October 2008. DOI: 10.1002/adma.200801167 107

[97] Cosmin Farcau, Helena Moreira, Benoit Viallet, Jeremie Grisolia, Diana Ciuculescu-Pradines, Catherine Amiens, and Laurence Ressier. Monolayered wires of gold colloidal nanoparticles for high-sensitivity strain sensing. *Journal of Physical Chemistry C*, 115(30):14494–14499, August 2011. DOI: 10.1021/jp202166s 107

[98] R. G. Freeman, K. C. Grabar, K. J. Allison, Bright R. M., J. A. Davis, A. P. Guthrie, M. B. Hommer, M. A. Jackson, P. C. Smith, D. G. Walter, and M. J. Natan. Self-assembled metal colloid monolayers: An approach to SERS substrates. *Science*, 267(5204):1629–1632, March 1995. DOI: 10.1126/science.267.5204.1629 107

[99] Yaroslav A. Urzhumov, Gennady Shvets, Jonathan A. Fan, Federico Capasso, Daniel Brandl, and Peter Nordlander. Plasmonic nanoclusters: A path towards negative-index metafluids. *Optics Express*, 15(21):14129, October 2007. DOI: 10.1364/oe.15.014129 108

[100] Jonathan A. Fan, Chihhui Wu, Kui Bao, Jiming Bao, Rizia Bardhan, Naomi J. Halas, Vinothan N. Manoharan, Peter Nordlander, Gennady Shvets, and Federico Capasso.

Self-assembled plasmonic nanoparticle clusters. *Science*, 328(5982):1135–1138, May 2010. DOI: 10.1126/science.1187949 108

[101] Jonathan A. Fan, Kui Bao, Li Sun, Jiming Bao, Vinothan N. Manoharan, Peter Nordlander, and Federico Capasso. Plasmonic mode engineering with templated self-assembled nanoclusters. *Nano Letters*, 12(10):5318–5324, October 2012. DOI: 10.1021/nl302650t 108, 109

[102] Valentin Flauraud, Massimo Mastrangeli, Gabriel D. Bernasconi, Jeremy Butet, Duncan T. L. Alexander, Elmira Shahrabi, Olivier J. F. Martin, and Juergen Brugger. Nanoscale topographical control of capillary assembly of nanoparticles. *Nature Nanotechnology*, 12(1):73–80, January 2017. DOI: 10.1038/nnano.2016.179 108, 109

[103] Ju Won Jeon, Petr A. Ledin, Jeffrey A. Geldmeier, James F. Ponder, Mahmoud A. Mahmoud, Mostafa El-Sayed, John R. Reynolds, and Vladimir V. Tsukruk. Electrically controlled plasmonic behavior of gold nanocube@Polyaniline nanostructures: Transparent plasmonic aggregates. *Chemistry of Materials*, 28(8):2868–2881, May 2016. DOI: 10.1021/acs.chemmater.6b00882 109

[104] B. Reiser, L. González-García, I. Kanelidis, J. H. M. Maurer, and T. Kraus. Gold nanorods with conjugated polymer ligands: Sintering-free conductive inks for printed electronics. *Chemical Science*, 7(7):4190–4196, June 2016. DOI: 10.1039/c6sc00142d 109

[105] Dhriti Nepal, M. Serdar Onses, Kyoungweon Park, Michael Jespersen, Christopher J. Thode, Paul F. Nealey, and Richard A. Vaia. Control over position, orientation, and spacing of arrays of gold nanorods using chemically nanopatterned surfaces and tailored particle-particle-surface interactions. *ACS Nano*, 6(6):5693–5701, June 2012. DOI: 10.1021/nn301824u 109

[106] Songbo Ni, Heiko Wolf, and Lucio Isa. Programmable assembly of hybrid nanoclusters. *Langmuir*, 34(7):2481–2488, February 2018. DOI: 10.1021/acs.langmuir.7b03944 109

[107] Michael Christian Gwinner, Elisabeth Koroknay, Fu Liwei, Piotr Patoka, Witold Kandulski, Michael Giersig, and Hamid Giessen. Periodic large-area metallic split-ring resonator metamaterial fabrication based on shadow nanosphere lithography. *Small*, 5(3):400–406, February 2009. DOI: 10.1002/smll.200800923 109

[108] A. Kosiorek, W. Kandulski, P. Chudzinski, K. Kempa, and M. Giersig. Shadow nanosphere lithography: Simulation and experiment. *Nano Letters*, 4(7):1359–1363, July 2004. DOI: 10.1021/nl049361t 109

[109] Yongdong Jin. Engineering plasmonic gold nanostructures and metamaterials for biosensing and nanomedicine. *Advanced Materials*, 24(38):5153–5165, October 2012. DOI: 10.1002/adma.201200622 109

[110] Alex Nemiroski, Mathieu Gonidec, Jerome M. Fox, Philip Jean-Remy, Evan Turnage, and George M. Whitesides. Engineering shadows to fabricate optical metasurfaces. *ACS Nano*, 8(11):11061–11070, November 2014. DOI: 10.1021/nn504214b 109

[111] Kai Chen, Thang Duy Dao, Satoshi Ishii, Masakazu Aono, and Tadaaki Nagao. Infrared aluminum metamaterial perfect absorbers for plasmon-enhanced infrared spectroscopy. *Advanced Functional Materials*, 25(42):6637–6643, November 2015. DOI: 10.1002/adfm.201501151 109

[112] Kai Chen, Bharath Bangalore Rajeeva, Zilong Wu, Michael Rukavina, Thang Duy Dao, Satoshi Ishii, Masakazu Aono, Tadaaki Nagao, and Yuebing Zheng. Moiré nanosphere Lithography. *ACS Nano*, 9(6):6031–6040, June 2015. DOI: 10.1021/acsnano.5b00978 109

[113] Guanqiao Zhang, Chuwen Lan, Huilong Bian, Rui Gao, and Ji Zhou. Flexible, all-dielectric metasurface fabricated via nanosphere lithography and its applications in sensing. *Optics Express*, 25(18):22038, September 2017. DOI: 10.1364/oe.25.022038 109

[114] You Zhou, Hanyu Zheng, Ivan I. Kravchenko, and Jason Valentine. Flat optics for image differentiation. *Nature Photonics*, 14(5):316–323, May 2020. DOI: 10.1038/s41566-020-0591-3 109

[115] Mohammadreza Khorasaninejad and Federico Capasso. Metalenses: Versatile multifunctional photonic components. *Science*, 358(6367):8100, December 2017. DOI: 10.1126/science.aam8100 109

[116] Guoxing Zheng, Holger Mühlenbernd, Mitchell Kenney, Guixin Li, Thomas Zentgraf, and Shuang Zhang. Metasurface holograms reaching 80% efficiency. *Nature Nanotechnology*, 10(4):308–312, April 2015. DOI: 10.1038/nnano.2015.2 109

[117] Jun Luo, Bo Zeng, Changtao Wang, Ping Gao, Kaipeng Liu, Mingbo Pu, Jinjin Jin, Zeyu Zhao, Xiong Li, Honglin Yu, and Xiangang Luo. Fabrication of anisotropically arrayed nano-slots metasurfaces using reflective plasmonic lithography. *Nanoscale*, 7(44):18805–18812, November 2015. DOI: 10.1039/c5nr05153c 110

[118] Liqin Liu, Xiaohu Zhang, Zeyu Zhao, Mingbo Pu, Ping Gao, Yunfei Luo, Jinjin Jin, Changtao Wang, and Xiangang Luo. Batch fabrication of metasurface holograms enabled by plasmonic cavity lithography. *Advanced Optical Materials*, 5(21):1700429, November 2017. DOI: 10.1002/adom.201700429 110

[119] R. Verre, M. Svedendahl, N. Odebo Länk, Z. J. Yang, G. Zengin, T. J. Antosiewicz, and M. Käll. Directional light extinction and emission in a metasurface of tilted plasmonic nanopillars. *Nano Letters*, 16(1):98–104, January 2016. DOI: 10.1021/acs.nanolett.5b03026 110

[120] Ming Lun Tseng, Yao Wei Huang, Min Kai Hsiao, Hsin Wei Huang, Hao Ming Chen, Yu Lim Chen, Cheng Hung Chu, Nien Nan Chu, You Je He, Chia Min Chang, Wei Chih Lin, Ding Wei Huang, Hai Pang Chiang, Ru Shi Liu, Greg Sun, and Din Ping Tsai. Fast fabrication of a Ag nanostructure substrate using the femtosecond laser for broad-band and tunable plasmonic enhancement. *ACS Nano*, 6(6):5190–5197, June 2012. DOI: 10.1021/nn300947n 110

[121] F. Watt, A. A. Bettiol, J. A. Van Kan, E. J. Teo, and M. B. H. Breese. Ion beam lithography and nanofabrication: A review. *International Journal of Nanoscience*, 04(03):269–286, June 2005. DOI: 10.1142/s0219581x05003139 110

[122] Alongkorn Pimpin and Werayut Srituravanich. Review on micro- and nanolithography techniques and their applications. *Engineering Journal*, 16(1):37–56, January 2012. DOI: 10.4186/ej.2012.16.1.37 110

[123] Seong-Jun Jeong, Ju Young Kim, Bong Hoon Kim, Hyoung-Seok Moon, and Sang Ouk Kim. Directed self-assembly of block copolymers for next generation nanolithography. *Materials Today*, 16(12):468–476, December 2013. DOI: 10.1016/j.mattod.2013.11.002 110

CHAPTER 5

Dynamic Metamaterial Absorbers

Having matured over the last two decades, there is continuing and growing interest in metamaterials for realization of applications. The ability to tune or dynamically control the novel responses exhibited by metamaterials would bolster this quest, thus ushering in the next revolution in materials and electromagnetic devices. Active control of metamaterials enables modulation of the properties of electromagnetic waves in amplitude, phase, polarization, and direction in real time, thus further extending their exotic properties, with the option of switching between resonant states. Metamaterial absorbers achieve a state consisting of critically coupled degenerative modes, and offer new strategies to further enhance the dynamic manipulation of electromagnetic waves. In this chapter, we will focus on perfect absorbers and give an overview of various approaches for achieving active control.

5.1 INTRODUCTION

Advanced technologies that have shaped our modern world, such as telecommunication, imaging, and data fusion are established on the ability to receive, tune and/or emit electromagnetic (EM) waves. Materials are the essential foundation which permits the control of EM waves thus enabling these technologies. However, the responses achievable from natural materials represent only a small fraction of the EM properties possible. Artificial EM materials, i.e., metamaterials, can be fashioned which permit full access to the theoretically possible response. The ability to modulate metamaterials in real-time has and will continue to enhance and shape the future technologies.

The interaction of EM waves with metamaterials creates highly localized and strong fields in a very small volume, such that a tiny variation in the resonator geometry, constituent materials, and/or surrounding environment, leads to a remarkable change of the resonant response, including frequency shift, amplitude change, phase, and polarization rotation. Through placement of functional media or elements at the field-concentrated regions, one may apply external stimuli, e.g., optical, temperature, voltage, or with magnetic field, to modify the material properties, thereby modulating the resonators EM response. These approaches may also be applied to metamaterial absorbers, whose perfect absorption condition can be easily breached by either changing the effective impedance matching condition or shifting the effective loss rate from the critical point. As described in Chapter 2, Eq. (2.41), the reflection coefficient for metal-

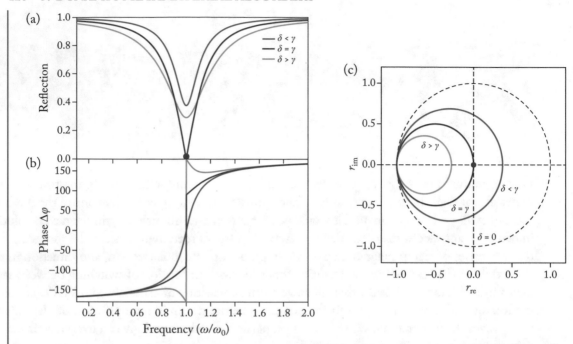

Figure 5.1: (a) and (b) Theoretically calculated reflection amplitude and phase spectra for metal based MPAs with three representative cases, where the metasurface in placed in the state of under-(blue), critical-(red), and overcoupling (green) based on temporal coupled-mode theory. (c) The corresponding complex reflection coefficient plotted in the complex plane. The dashed circle presents reflection for lossless resonating systems. The red dot indicates the reflection minimum under the critical coupled condition.

dielectric-metal absorbers is dependent on the radiative loss rate γ and material loss rate δ. The specific dependence is

$$r_{\text{GPA}} = \frac{i(\omega - \omega_0) - (\delta - \gamma)}{-i(\omega - \omega_0) + (\delta + \gamma)}. \tag{5.1}$$

The radiative loss rate γ is set by the geometry of the metamaterial unit cell and the periodicity, whereas the material loss δ rate is given by the constituent materials. Beginning from zero (or low) material loss, the resonator system operates in an under-coupled state, i.e., $\delta < \gamma$. Here the minimum in the reflectivity is relatively high, meaning low absorptivity, and the bandwidth is relatively narrow—see blue curves of Fig. 5.1. As material loss is increased, we reach a state of critical coupling $\delta = \gamma$, and the reflectivity goes to zero, whereas the reflected phase undergoes an abrupt change at ω_0 from $-90°$ at low frequency, to $+90°$ at higher frequency. A further increase of the material loss rate brings us to a state of overcoupling $\delta > \gamma$ and the minimum in reflectivity at ω_0 increases from zero, and has a relatively large bandwidth. Surprisingly, the phase change can be over 180° and up to 360°, which is much larger than that a simple resonating system can achieve. Furthermore, the complex reflection coefficient can be further separated into its real and imaginary parts and satisfies [1–3]:

$$\left(r_1 + \frac{\delta}{\gamma + \delta} \right)^2 + r_2^2 = \left(\frac{\gamma}{\gamma + \delta} \right)^2, \tag{5.2}$$

where r_1 and r_2 are the real and imaginary parts of the complex reflectivity coefficient. Clearly, the real part and imaginary part of reflection coefficient changes along a circle in a complex plane, centered on the real axis at $(\delta/(\gamma + \delta), 0)$, with a radius of $\gamma/(\gamma + \delta)$. For a passive system, since both γ and δ are equal or larger than zero, the circle center is on the left half-plane and always crosses $(-1, 0)$. Figure 5.1c shows the complex reflection coefficient, where the curve is parameterized by frequency, with color corresponding to that in Fig. 5.1a. When the system is in the critically coupled condition, the reflection crosses the origin, which corresponding to zero reflection as shown in Fig. 5.1a. In the overcoupling state, the reflection circle shrinks to the left negative plane, and the phase covers the range from 90–270°. However, for the under-coupling condition, the reflection reaches all four quadrants, such that the maximum reflected phase covers from 0–360°. The takeaway from all of the above analysis and discussion is that we observe that the reflection can be easily manipulated by modifying material loss in MPAs. We should mention that active control of MPAs is not limited to only the modulation of δ. Studies have shown that it is also possible to achieve tunable response by changing the resonant frequency [4, 5], radiative loss [6, 7], and even cross-coupling between two coupled modes [8]. In the following sections, we will overview the efforts on tunable metamaterial absorbers which categorized by the tuning mechanisms, i.e., electric, optical, mechanical, thermal, and magnetic tuning.

5.2   ELECTRICALLY TUNABLE METAMATERIAL ABSORBERS

The maturity of both the electronics and semiconductor industries enables the integration of high performance components within metamaterial unit cells and, in many cases, permits the realization of real-time, high-speed, broad bandwidth, high dynamic range variable electromagnetic properties for practical applications, including imaging and beam steering [9]. In contrast to conventional metallic metasurfaces where an extra electrode has to be designed for applying bias voltage, the metal-dielectric-metal configuration of MPAs permits the application of a bias directly applied between the ground and the top metamaterial portions. To date, electrically tunable MPAs have been successfully demonstrated from radio frequencies (RFs), and microwave to infrared regions.

The first electrical tuning of MPA was demonstrated in the microwave frequencies [10]. Compared to other frequency bands, electrical tuning of microwave MPAs is significantly easier, since active lumped elements can be directly placed in the metamaterial array [11–14]. The

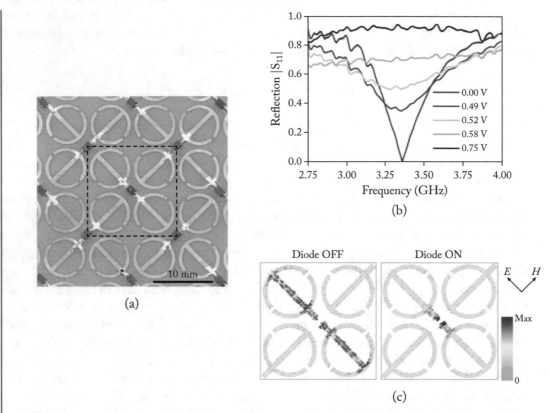

Figure 5.2: Electrically tunable MPAs using lumped elements. (a) Image of fabricated the switchable microwave metamaterial between total reflection and absorption states. ELC resonators are connected with diodes. The dashed square indicates one unit cell of metamaterial with four ELCs. (b) Measured reflection spectra under normal incidence with various forward bias voltages. (c) Surface current density distribution on ELCs at peak absorption frequency under the same polarized incidence. Reprinted with permission from Ref. [15] © 2010 AIP.

MPA unit-cell consists of 2 × 2 circular ELC resonators with orthogonal orientation between neighbors, fabricated on a copper supported FR4 substrate. Arrayed diodes were placed between two ELC resonators with the same orientation, as shown in Fig. 5.2a, so that the bias voltage was applied between the top metamaterials and bottom ground plane through vias. When a zero bias voltage is applied on the diodes (OFF state), the diode is capacitive so that opposite charges accumulate on the arches with a strong resonance. As a forward bias voltage is applied, the diode becomes inductive and less capacitive, leading to a significant resonance shift to higher frequencies. Figure 5.2a shows the fabricated metamaterial sample based on the optimized design by integrating the metallic structures on a FR4 dielectric substrate using standard print circuit board (PCB) technique. Figure 5.2b shows the measured reflection coefficient spectra as a function of applied voltage. At zero bias, a nearly perfect absorption was achieved at 3.34 GHz with a FWHM of 17%, leading to a strong resonating surface current on those ELCs with central bar parallel to the polarization as shown in Fig. 5.2c. When the forward bias was increased to 0.75 V, the reflection amplitude went back to nearly unity with negligible surface current in the ELCs. Since the orthogonally oriented elements can be tuned independently, the polarization of reflected wave can be switched from a linear polarization to an elliptical polarization [11]. Similarly, the reflection/absorption conversion can also be obtain by connecting neighboring metamaterial elements in series using diodes [12].

At higher frequencies, as mentioned previously, direct insertion of lumped elements in the metamaterial array is impractical because of the decreased unit-cell size, and fabricational difficulties. Nonetheless, several other methods have been utilized to overcome the lack of commercially available lumped element devices. One particular intriguing way involves the integration of MPAs with semiconductors, 2D materials, and some functional material like liquid crystals (LCs) and transparent conductive oxides (TCOs), which have been shown as ideal materials for tuning metamaterial response in the terahertz, infrared regimes. Liquid crystals have been known to exhibit voltage dependent birefringence and prevailed in the optical display market over the past several decades. Because of capacitive loading between the top metamaterial and bottom ground plane, replacing the dielectric spacer with LCs allows for a voltage dependent spacer refractive index, resulting in a resonant frequency shift [16–18]. The liquid crystal 4'n-pentyl-4-cyanobiphenyl (5CB) was initially used for the experimental demonstration of LC based tunable MPA at terahertz frequencies [16]. As shown in Fig. 5.3a, the ELC resonators are supported by patterned polyimide spacer, where the 5CB is filled and encapsulated. The inset of Fig. 5.3b shows an optical microscopic image of a 2 × 2 metamaterial array. With no bias voltage applied, a resonant absorptivity with peak value of 85% and a FWHM of 600 GHz at 2.62 THz was obtained. As an AC voltage $V_{bias} = 4$ V with frequency of 1 kHz was applied between the top resonators and ground plane, the resonant frequency shifted to 2.5 THz, with a little lower absorption peak value of 80% and a narrower FWHM of 420 GHz. A modulation depth of 35% from the resonant frequency shift, and 4.6% resonant frequency shift were obtained. Further arrangement of the LC tunable MPAs into a 6 × 6 pixel array demonstrated that each pixel can be controlled independently, highlighting the great potential for THz applications, such as imaging and beam steering [17].

LC-based tunable MPAs can be easily implemented using modern LC display technologies, however, small modulation bandwidth around 100 kHz [16] impedes them for high-speed and large-bandwidth applications, such as communication and computational imaging. Through electrical doping or depleting charge carriers inside materials, modulation speed can be up to hundreds of GHz. Driven by external voltages, the carrier density in normal bulk semiconductors, two-dimensional electron gas (2DEG) systems, TCOs, and 2D materials can be altered by several orders so that these materials are switched between plasmonic and dielectric

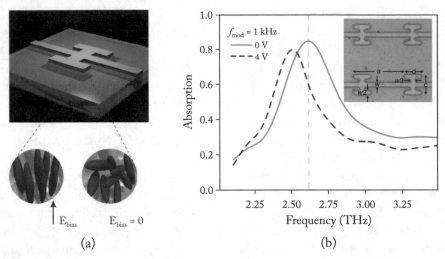

Figure 5.3: Liquid crystal-based tunable MPAs. (a) Schematic of a single unit cell of the liquid crystal metamaterial absorber. The red lines point to the depiction of the random alignment of liquid crystal in the unbiased case (right) and for an applied ac bias (left). (b) Experimentally measured absorptivity spectra for 0 V (blue solid curve) and 4 V (red dashed curve) at modulation frequency of 1 kHz. Dashed line is centered at $A_{max}(V_{bias} = 0) = 2.62$ THz. The inset shows the optical microscopic image of a portion of the metamaterial array where the dimensions of the unit cell are: $a = 50$, $c = 20$, $d = 16$, $w2 = 4.5$, and $w3 = 5$ in micrometers. Reprinted with permission from Ref. [16] © 2013 APS.

states with various conductivity. At terahertz frequencies, metallic metamaterial arrays can be directly fabricated on doped semiconductor thin film, forming a Schottky junction [20], where an Ohmic contact is established between the semiconductor and bottom ground plane. As shown in Fig 5.4a, the MPA consists of two metallic layers with a 2-μm dielectric spacer, i.e., a doped epitaxial layer of GaAs with 2×10^{16} cm^{-3}, lying in-between. The bottom ground plane is electrically connected to electrodes via indium bumps. Through changing the top metamaterial geometries, four pixels operating at different frequencies were fabricated, as shown in Fig. 5.4b. As a reverse bias voltage is applied between the metamaterial and semiconductor, the carriers in the contact regions will be gradually depleted. A maximum $V_{bias} = -26.5$ V leads to a thorough depletion of electrons in 2-μm n-type GaAs layer to shift the frequency of the reflectance minimum by nearly 5% for the pixel designed at 2.72 THz. The average modulation depth near resonant frequencies for four pixels is about 54%. Figure 5.4c plots the reflection at resonant frequency $R(\omega_0)$ as a function of V_{bias} for each color pixel.

Experiments showed that the modulation depth can be up to about 30% with modulation speed in the MHz range, which is limited by the semiconductor mobility and the de-

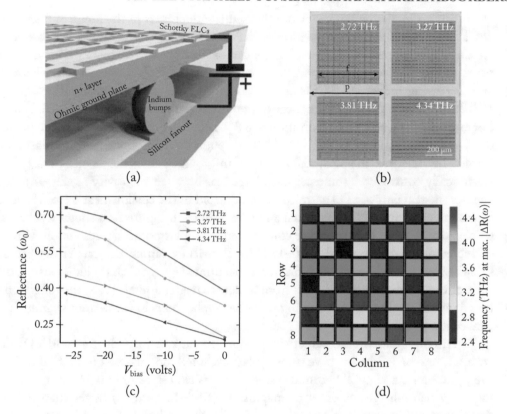

Figure 5.4: Electronically controlled semiconductor based MPAs (a) Cross-sectional schematic view of the single pixel of an electronically controlled THz metamaterial absorber with doped GaAs as the spacer for dielectric properties modulation. (b) Optical microscopic images for each color pixel at 2.72 THz, 3.27 THz, 3.81 THz, and 4.34 THz, respectively. (c) Reflectance at resonant frequency ω_0 as a function of bias (V_{bias}) for each color subpixel. (d) Spatial light modulator performance with the corresponding frequency for each pixel with maximum change in reflectance for applied bias at 0 V and −26.5 V. Reprinted with permission from Ref. [19] © 2013 Wiley-VCH.

vice area accompanied by a high stray capacitance or parasitic capacitance from electrodes and wires [9, 21, 22]. Utilization of such technology, an 8 × 8 pixelated spatial light modulator allows terahertz single-pixel imaging with high-frame-rate and high-fidelity images. In addition to the tuning based on Schottky junctions, it has also been shown that metallic metasurfaces fabricated on multi-quantum-well (MQW) semiconductor heterostructures can polaritonically couple to intersubband transition for large quantum-confined Stark effect, which is rapidly modulated by the applied bias between the top metasurface and bottom ground plane [23].

Since the plasma frequency of most semiconductors is in the range of mid-wave infrared and above [24], electrical modulation of MPAs at shorter wavelength becomes challenging. As an alternative to semiconductors, TCOs can exhibit plasma frequencies in the near-infrared range [25–28]. These oxides are widely used as transparent electrodes in solar energy harvesting and displaying industries. TCOs can be doped to have carrier concentrations between 10^{19}–10^{21} cm^{-3} by manipulating the concentration of oxygen vacancies and interstitial metal dopants. For example, for electron concentration of 8.8×10^{19} cm^{-3}, the fitted plasma frequency based on the Drude model, is $\omega_p \approx 139$ THz [29]. Through inserting the ITO into plasmonic cavities, formed between the metamaterial and bottom ground plane, the carrier density can be electrostatically controlled by an external voltage bias between the cavity [3, 29–32]. This method enables the real part of ITO permittivity modulated in the epsilon-near-zero frequency regime, thus achieving large reflectance change up to 15% [30]. Figure 5.5a shows a schematic plot of an electrically tunable metasurface constructed from plasmonic gap resonator, which is capable of achieving phase change from 0–180° by electrically tuning the carrier density of ITO inside the gap, where the field significantly concentrated (Fig. 5.5c). By judiciously designing and controlling the resonant properties from under- to overcoupling states, the reflection phase was tuned over 180° (Fig. 5.5d), allowing potential applications in wave-front shaping, polarization conversion [3].

Attempts based on conductivity change on bulk semiconductors and 2DEG materials with limited carrier doping have shown relatively small modulation on the EM response. However, graphene, a truly 2D atomic system with thickness of only 0.345 nm, allows the doping from neutrality to an order of magnitude (10^{13}–10^{14} cm^{-2}) larger than that of conventional semiconductors. The optical conductivity of graphene, derived based on the random-phase approximation [33, 34], is attributed to the intraband transitions and interband transitions, which normally take place in the infrared and terahertz ranges [31, 35]. In the terahertz range, intraband-transition dominant graphene, behaving like a Drude-type semiconductor, has been shown to exhibit resonant plasmonic response [35, 36]. When hybridized with MPAs with locating the graphene layer underneath the metamaterial structure, experimental demonstration showed that nearly 2π phase change is possible to achieve through electrostatic modulation of the carrier depletion or accumulation in the graphene at terahertz frequencies. Figure 5.6a shows a schematic view of a graphene based tunable MPA operating at terahertz frequencies. The MPA consists of a 5-layer stack, including an aluminum ground plane, a layer of 85-μm-thick SU-8 spacer, an array of Al mesas with size of 100 μm $\times$ 160 μm (Fig. 5.6b). To achieve dynamic modulation, a layer of graphene was transferred onto the structure and subsequently covered by a layer of ion-gel superstrate as a top gate for applying a bias voltage to tune the resistance of graphene. Figure 5.6c–d display the measured phase change for various gate voltages relative to the charge neutral point ($\Delta V_g = V_g - V_{CNP}$). As the gate voltage increases from zero, the amplitude of reflection drops continuously to minimum value with $\Delta V_g = 0.76$ V. Clearly, the phase variation across the frequencies decreases from 360–180°, indicating a tran-

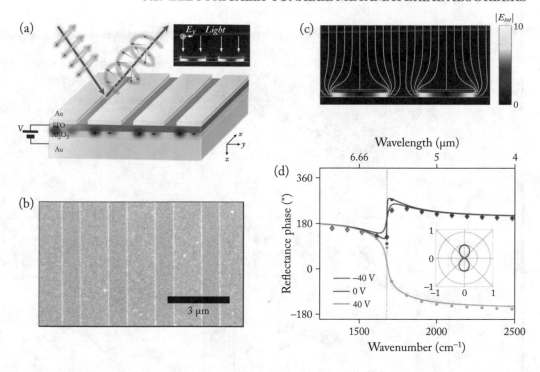

Figure 5.5: Electronically controlled semiconductor based MPAs. (a) Illustration of the config-uration of an active metasurface capable of achieving a 180° phase change in the reflection. The phase change is controlled by electrically tuning the carrier density in ITO. The inset shows the electric fields near the metasurface on resonance. (b) An SEM image of fabricated meta-surface with 1400-nm-wide stripes. (c) Electric field images and incident power flow lines for illumination of a metasurface with 1400-nm-wide stripes at wavelength 6 μm. (d) Measured spectral dependence of the reflection phase of light for three different bias voltages in a state of under-coupling (blue), near critical-coupling (red), and overcoupling (green). An electrically induced phase change of 180° can be achieved at a wavenumber of 1680 cm^{-1} (gray vertical line). Reprinted with permission from Ref. [3] © 2017 ACS.

sition from an under-coupled state to a critical-coupled state. A further increase in the gate bias beyond the critical voltage, the phase variation only changes within 180° due to the much more conductive graphene layer which dampens the resonance—consistent to previous analysis showing an over-coupled MPA. Overall, the measured results demonstrated an absolute phase modulation of ±180° associated with the gate at frequencies around the resonance [2]. Based on such a modulation scheme, various optical devices such as polarization modulators [2, 37] and absorber/reflector converter [38] operating in the terahertz range were proposed and achieved.

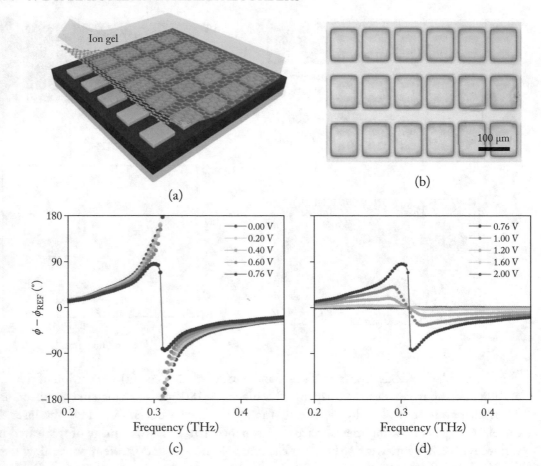

Figure 5.6: Tunable graphene-based terahertz MPAs. (a) A schematic of a tunable graphene based MPA operating at terahertz frequencies. The metasurface consists of an array of aluminum mesas and a continuous Al film separated by an SU-8 spacer. CVD grown graphene is transferred to the top of the metasurface and covered by a layer of ion-liquid gel. The gate voltage is applied between the graphene and ion-liquid gel. (b) An optical microscopic image of the fabricated metamaterial modulator. (c), (d) Measured reflection phase with various gate voltages ΔV_g. Reprinted with permission from Ref. [2] © 2015 APS.

Graphene-based electrically tunable MPAs have also been demonstrated in the mid-infrared region [5, 38–46]. As shown in Fig. 5.7a, the top metasurface layer is separated from a back aluminum ground plane by a thin oxide film. Because of its ultra thin thickness compared to its operational wavelength, a graphene layer can be treated as a 2D layer with an effective surface conductivity for analytically designing EM response of metamaterials [39, 47]. As the gate

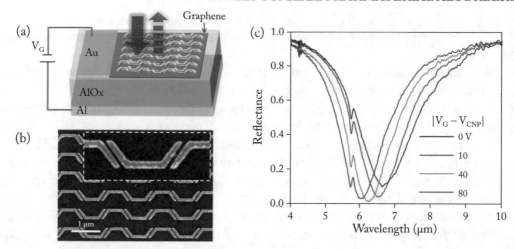

Figure 5.7: Electrically tunable graphene based MPAs. (a) Schematic of the ultrathin optical modulator based on a tunable metasurface absorber. (b) An SEM image of the metasurface on graphene. The inset shows a close-up view of a portion of the metasurface. (c) Measured reflection spectra (normalized to the reflection spectrum of an aluminum mirror) of the MPA in (a) at different gate voltages $|V_G - V_{CNP}|$, where V_{CNP} is the gate voltage when the concentrations of electrons and holes in the graphene sheet are equal, i.e., charge neutral point (CNP). Reprinted with permission from Ref. [39] © 2014 ACS.

voltage is tuned away from the charge neutral point, where the concentrations of electrons and holes are the same in the graphene with minimum conductivity, the imaginary part of graphene sheet conductivity increases while the real part increases less [5]. As a result, the absorption peak blue-shifts to shorter wavelength. Experimental measurements showed the achieved maximum modulation depth is about 80% [5, 46]. Due to the electrostatic doping, the modulation speed is only determined by the RC constant of the circuit, where R is the sum of the graphene sheet resistance, contact resistance, and load resistance and C is the total capacitance including the device capacitance between graphene and ground plane, capacitance from electrodes, and parasitic capacitance from wires. If the size of an MPA device is made small, the modulation speed can be increased and GHz modulation is possible [46, 48].

We have highlighted a number of electrically tunable metamaterial devices, some of which have demonstrated efficient modulation and offer the advantage of integrated circuit compatibility, low-cost, real-time addressability, and high-speed operation. The subwavelength nature of metamaterials further enables the fabrication of compact and low-power devices, which will support the realization of commercially viable miniaturized sensors, detectors and modulators.

5.3   MECHANICALLY TUNABLE METAMATERIAL ABSORBERS

Metamaterials described in the previous section have demonstrated electrical tuning by modification of substrate properties. Alternatively, it is possible to change the physical metamaterial parameters, such as the lattice [49, 50], reshaping of elements [51], and rotating elements [52], to modify the local dielectric environment and hence the resonant response. For example, if metamaterials are incorporated into microelectromechanical systems (MEMS) or nanoelectromechanical systems (NEMS), tunability may be achieved by mechanical actuation [53–55]. When integrating MEMS and NEMS technologies with metal based metamaterial absorbers, both of the metal layers, i.e., the patterned layer and the bottom ground plane, can be directly implemented as electrodes for application of voltage for electrostatic actuation and mechanical deformation. Over the past years, a variety of mechanically reconfigurable metamaterial absorbers have been experimentally demonstrated from terahertz to near infrared wavelengths, and have exhibited large EM modulation depth, ease of integration with integrated circuits, fast response and polarization selectivity. Some have demonstrated use for practical applications, such as terahertz imaging [56, 57], thermal emission control [58], and mechanical oscillation control [59].

Split-ring resonators fabricated on bimaterial cantilevers have been shown to enhance far-infrared detection with local heating led by the absorption. However, as only one resonant mode can be supported by the resonator, at most 50% of the incident power from one side can be absorbed. Therefore, to further increase the responsivity, Alves et al. experimentally demonstrated a terahertz sensor integrating a MPA with bimaterial cantilevers [56]. A bimaterial cantilever normally consists of two layers of thin films with drastically different thermal expansion coefficients. The beam deflection deflection with corresponding temperature change ΔT is given as [60]

$$\Delta z = \frac{3E_1 E_2 t_1 t_2 (t_1 + t_2) L^2}{(E_1 t_1)^2 + (E_2 t_2)^2 + 2E_1 E_2 t_1 t_2 (2t_1^2 + 3t_1 t_2 + t_2^2)} (\alpha_2 - \alpha_1) \Delta T, \qquad (5.3)$$

where E_1 and E_2 are Young's moduli of the two films; t_1, t_2, and α_1, α_2 are the thickness and linear thermal expansion coefficients of the two bimorph layers; L is the length of the cantilever. As shown in Fig. 5.8, the terahertz detector consists of Al squares with size of 16.5 μm and periodicity of 20 μm and an Al ground plane separated by a 1.2 μm-thick SiO_2 layer. The whole absorber converts the absorbed terahertz radiation into heat such that the bimaterial cantilevers bend away from the original position due to the thermal gradient induced stress inside cantilevers. As a result, the responsivity can be optically defined as a function of rotation angles through a reflected optical beam incident on the metamaterial layer. The measured absorptivity spectrum exhibited an absorption peak of 90% at 3.8 THz, which coincides with the operating frequency of a terahertz QCL source. The detector was characterized with a responsivity of 0.1 deg/μW and a time constant of 14 ms.

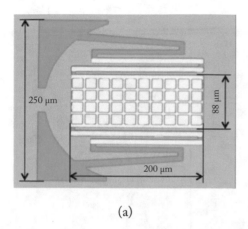

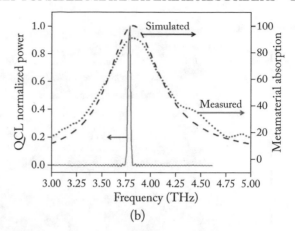

(a)   (b)

Figure 5.8: Terahertz detector based on metamaterial absorber. (a) Microscopic image of the bimaterial MEMS sensor with patterned square-shape metamaterials before releasing. (b) Measured (dotted curve) and simulated (dashed curve) absorption spectra of the metamaterial absorber. The peak absorption matches well with the measured QCL emission frequency (solid red line). Reprinted with permission from Ref. [56] © 2012 OSA.

The detector shown above is passive with the mechanical deformation relying on the existence of terahertz source. However, to dynamically modulate EM waves, an active device with mechanical deformation driven by external stimuli, such as voltage, heat, is required. In 2013, Liu and his colleagues, for the first time, experimentally demonstrated a dynamic infrared MPA which is based on an MEMS parallel-plate actuator [4]. In contrast to other absorber designs, here the top Babinet metamaterial layer and bottom ground plane were separated by an air gap. The top metal layer and its underline dielectric (MgF_2) layer are suspended on beams at four corners, as shown in Fig. 5.9a,c. Without applying voltage between two metal layers, the metamaterial layer is about 1.2 μm away from the ground plane so that the metamaterial is not in the critical-coupling state. As the applied voltage larger than 10 V, i.e., pull-in voltage, the whole metamaterial layer was snapped down to the bottom to reach the perfect absorption state. Experimental characterization using an infrared microscope with a Fourier transform infrared (FTIR) spectrometer showed the achieved modulation depth on the absorptivity was as large as 56% at a wavelength of 6.2 μm. The whole device can be modulated in the kHz range with a 3dB point of 30 kHz. Similar reconfigurable MEMS absorber based on Babinet metamaterials was also scaled to operate in the near-infrared range [61]. As the device modulates the absorptivity in a real-time mode, equivalently, it can also be considered as a reconfigurable emitter with dynamic control on emissivity spectra. Not like conventional demonstrated systems, which are either based on a change in the temperature or induced by a specific property of a natural material, such as phase change material, an MEMS-based reconfigurable MPA can be operated at

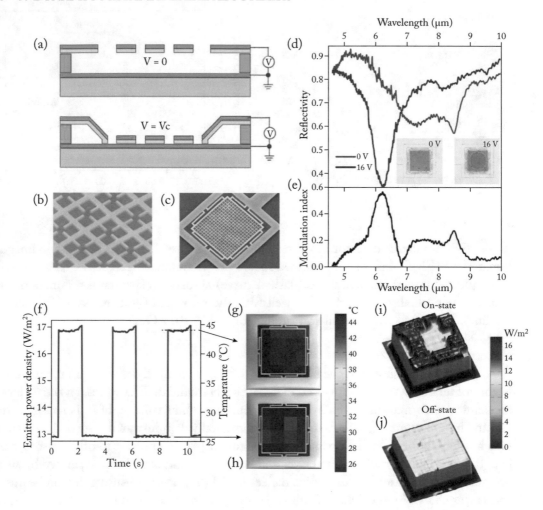

Figure 5.9: Dynamic metamaterial absorber based on MEMS actuators. (a) Schematic of the tuning mechanism of the metamaterial absorber with patterned electrostatic parallel-plate actuator in (top panel) open (bottom panel) snap-down status. (b) A close-up SEM image of the Babinet metamaterial structure. (c) An SEM perspective image of the metamaterial device supported with eight beams. (d) Experimental reflectivity from the MEMS MPA in the open (blue curve) and snap-down (red curve) status. (e) Modulation depth of the reflectivity. (f) Time dependent of the spatially averaged emitted power density (left y axis) and corresponding estimated temperature (right y axis). (g), (h) Spatial dependence of temperature over the pixel in on (g) and off (h) states. (i), (j) Emitted power density of MEMS metamaterial emitter with an IR image in (i) on and (j) off states. Reprinted with permission (a)–(e) from Ref. [4] © 2013 Wiley-VCH, (f)–(j) from Ref. [58] © 2017 OSA.

room temperature with a large modulation depth and a high modulation speed. Through scaling the geometry demonstrated in [4], Liu et al. experimentally showed a reconfigurable MEMS metamaterial emitter with emissivity peak of 0.95 at 8.9/μm. Upon application of bias voltages, the differential emissivity achieved a sharp peak with a value close to 70% at the resonant wavelength. Therefore, the spectral radiant existence detected by an IR camera can be calculated as:

$$M_\lambda(\lambda, T_R) = S(\lambda)E(\lambda)B(\lambda, T_R), \tag{5.4}$$

where $S(\lambda)$ is the responsivity of the IR camera, T_R is the environment temperature, $E(\lambda)$ is the spectral emissivity of the device, and $B(\lambda, T_R)$ is the spectral radiance of the blackbody. The total radiosity (emitted power density) change from the reconfigurable emitter is from 46 Wm^{-2} in the off state to a value of about 60 Wm^{-2} in the on state. Figure 5.9f shows measured spatially averaged emitted power density (left y axis) and temperature (right y axis) modulated at 0.5 Hz. The spatially averaged temperature changed from 25.4–44.3°C—a difference of 18.9°C—see Fig. 5.9g–h [58]. Such a temperature change is a result of the dynamic tailoring of spectral emissivity without a change in working temperature. Pixelation of such a device into a spatial light modulator with 8 × 8 pixels has been demonstrated for image displaying, as shown in Fig. 5.9.

There are still some significant challenges, such as MEMS structural reliability, stiction, electrostatic discharge, particulate contamination, etc., which may limit their commercialization as state-of-the-art devices. However, compared to electrical tunable metamaterials, dynamic tuning based on reconfigurable MEMS structures provide large modulation depth, broadband bandwidth, and more flexible modulation. MEMS-based tunable metamaterials will play an important future role in the development of dynamic tunable metamaterial devices, especially in the infrared range.

5.4   OPTICALLY TUNABLE METAMATERIAL ABSORBERS

As mentioned in Section 5.2, modulation of carrier concentration could induce a change of refractive index and material loss, thus leading to the absorption shifting among under-coupled, critical-coupled, and over-coupled states. Free carriers consisting of electrons and holes can also be generated in semiconductors when light of a critical energy is incident. In particular, this occurs when the photon energy is equal to or greater than the bandgap energy of the semiconductor. Such photogenerated carriers exist in the semiconductor for a period of time, determined by the carrier recombination time, and provide a metallic response that can be described by a Drude model and quantified by the plasma frequency. For frequencies below the plasma frequency the semiconductor behaves as a metal, while above it exhibits insulating properties. To date, metamaterials hybridized with semiconductors have been widely used for all-optical modulation of waves [62–66]. One distinct advantage of such an integrated device is the possibility of realizing ultrafast electromagnetic wave modulation with femtosecond optical pulses. For a sandwiched fishnet structure with an α-silicon spacer, a fast switching with response time of several picoseconds was achieved due to the fast carrier recombination [67].

The combination of MPA with semiconductors enables designs which may be realistically implemented in commercial fabricational processes. However, obtaining dynamical properties of metamaterial with optical control is significantly easier, can achieve higher modulation speeds, and can be used as a test bed for experimental realization of concepts. According to perturbation theory, a larger change in resonance frequency is possible, and more efficiently, when active media are located directly in field enhanced regions. When the gaps of split-ring resonators are filled with silicon thin film, fabricated on gold ground plane supported by a sapphire spacer, numerical analysis showed a modulation depth of about 60% at 1.1 THz, when the silicon conductivity changes by 3 orders [69]. Because of the significant conductivity change, the real part of the permittivity gradually becomes negative and the imaginary part of the permittivity increases by orders of magnitude in a broad range. This permits both modulation in the fundamental LC resonance, and also tuning of higher order resonances. Figure 5.10a shows a schematic of an optically tunable terahertz absorber with the entire ELC absorber fabricated on a sapphire substrate [68]. Through application of an optical pump beam, up to about 200 mW, a dual-band tunable absorber was demonstrated, achieving a modulation depth of 38% and 91% at the LC resonance of 0.7 THz and dipole resonance of 1.1 THz, respectively, as shown in Fig. 5.10c–d. Since the terahertz pulses are incident on the thick sapphire substrate, multiple reflections at the interface between the substrate and free space limited the actual absorbed power, and the resulting aliased pulse trains complicated signal processing. To avoid this problem, a free-standing MPA was later demonstrated through transferring patterned GaAs patches and metamaterial resonators onto a flexible polyimide spacer [70]. With only 25.6 mW of optical pumping power, the modulation depth of 25% and 40% were obtained at 0.78 THz and 1.75 THz, respectively.

Modulation of the absorbing response is often achieved through a tuning of the quasi-electric dipole response, but it is also possible to modify the quasi-magnetic dipole as well. As described in Chapter 2, arrayed dielectric disks may be designed to support two orthogonal dipole-like modes—an electrical (even) mode EH_{111}, and a magnetic (odd) mode HE_{111}. These two modes can be made degenerate via tuning the geometry, such as height, radius, and periodicity, thereby achieving Huygens' metasurfaces. However, through photoexcitation of carriers in the all-dielectric metasurface, the introduced material loss can break the degeneracy between the even mode and odd mode, subsequently modifying the metasurface absorptivity and transmitted phase. An optical-pump terahertz probe experiment on an array of silicon cylinders showed that an optical pulse with a fluence of $10\,\mu Jcm^{-2}$ can lead to a transmission modulation depth of 99.93% at the resonant frequency of 1.03 THz, and an associated phase change larger than $\pi/2$. Further numerical analysis determined the dependence of the even and odd modes with respect to material loss. As the top of silicon disks are photodoped by an 800-nm pulse with increasing fluence, both the resonant frequencies for even and odd modes increase while the odd eigenmode shows a strong dependence on the optical fluence in comparison to the even

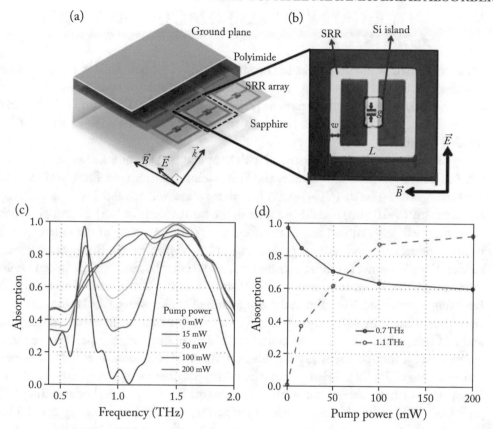

Figure 5.10: Optically tunable terahertz metamaterial absorber. (a) Schematic of the tunable absorber consisting of a gold ground plane, polyimide spacer, and gold SRR array on top of a sapphire substrate. (b) An optical microscope image of the unit cell of a fabricated device showing an ELC with a Si island in the capacitive gap. (c) Measured absorption spectra for various optical pump power and (d) the change in absorptivity as a function of pump power at 0.7 (black) and 1.1 (red) THz. Reprinted with permission from Ref. [68] © 2014 Wiley-VCH.

eigenmode. The odd eigenmode shows a dependence that is 2.5× greater than that of the even mode. As a result, the degeneracy of the Huygens' metasurface is lifted and the electromagnetic response changes remarkably.

5.5   THERMALLY TUNABLE METAMATERIAL ABSORBERS

Many materials undergo a drastic change of their electrical and/or electromagnetic properties as a function of temperature, including metal-to-insulator transitions [71], refractive index changes [72], carrier density variations [73], and structural deformation [60, 74]. Hybridization of metamaterial absorbers with thermally responsive materials enables a means for producing tunable response.

One relatively common material with temperature dependence are phase-change materials (PCMs), such as vanadium dioxide, and Germanium-antimony-tellurium (GST) chalcogenide glass. Vanadium dioxide (VO_2) exhibits a phase transition around 340 K enabled by the percolation of tiny (5–10-nm) metallic puddles in the insulating host [75]. During its phase transition from monoclinic to rutile phase, the effective medium response of the entire thin film, a composition of metallic and insulate puddles, can be described by the Bruggeman effective medium model. For a 75-nm VO_2 film deposited on a sapphire substrate, the measured far-infrared conductivity σ_1 can increase to about 5000 $(\Omega cm)^{-1}$ as the temperature increases to 380 K [76]. Inserting a patterned VO_2 film in the gap region of a microwave ELC resonator, which was fabricated on a sapphire substrate supported by a copper metal, realizes high absorption peaks at 9 and 18 GHz, which gradually weakened as the temperature was ramped over the transition temperature [77]. Similar thermally tunable metamaterial absorbers were also demonstrated in the infrared range [78, 79]. Arising from the nontrivial phase shifts at interfaces of an asymmetric Fabry–Perot cavity, perfect absorption was also achieved via directly depositing VO_2 thin film on a highly reflective substrate with no patterning. Through controlling the thickness of lossy thin film, the total interfered reflection at the interface, between air and the thin film, can reach zero, yielding a perfectly absorbing state. Due to the temperature-dependent permittivity of VO_2 with an insulator-to-metal transition, experiments showed a broadband reflection tuning with peak change from zero to about 80% for an ultra-thin ($\lambda/60$) film of VO_2 on a sapphire substrate [78].

Chalcogenide semiconductors, such as GST, are another type of phase-change materials which exhibit a transition between amorphous and crystalline phases. Compared to vanadium dioxide, such a phase transition is reversible without constant heating necessary to preserve the changed phase. A phase transition in chalcogenide glass can be induced by external heating or triggered by ultrafast optical pulses in a photothermal process. Because of different crystal structure between two phases, chalcogenide glass shows a remarkable change of permittivity and material loss tangent [81]. The chalcogenide glass has been widely used in rewritable DVDs and blue-ray disks and may be an option for the future non-volatile memory industry. Recently, the combination of metamaterials with chalcogenide glass has attracted considerable attention for tunable electromagnetic response. Likewise, chalcogenide glass was also implemented in metamaterial absorbers as a spacer to actively modify the absorptivity or emissivity in the infrared ranges [80]. Figure 5.11a shows a thermally tunable mid-infrared absorber with an array of

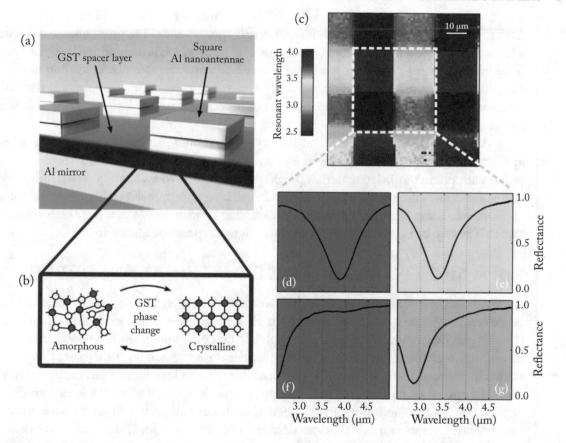

Figure 5.11: Thermally tunable infrared absorber using GST. (a) Illustration of the tunable absorber device. (b) Phase transition of GST from amorphous to crystalline state. (c) Pixelated four color-coded plasmonic absorbers with resonant wavelength from 2.5–4 μm. The Al nanoantenna length in these pixels is changed from 300–450-nm in a 50-nm step. (d–g) Measured reflectance spectra of the four pixels. The GST spacer layer is in the amorphous phase for all measurements. Reprinted with permission from Ref. [80] © 2015 Wiley-VCH.

square aluminum nanoantennas fabricated on a chalcogenide glass $Ge_2Sb_2Te_6$ (GST-236) thin film, with an aluminum ground plane. Due to the large refractive index of GST-236 in the mid-infrared range, the deposited GST spacer is 85-nm, which is only about $1/41\lambda$ for the GST in crystalline phase. As the temperature increased, the GST thin film underwent a phase transition from the amorphous to the crystalline phase, while the corresponding refractive index changed gradually from 3.5–6.5. As a result, the measured resonance redshifted by 0.7 μm and reflectance showed a significant increase by 60% at initial resonance. Since the resonance is mainly determined by the geometry, multiple patch resonators with various lengths can be tessellated into complicate patterns for multispectral imaging. Figure 5.11c shows a color-coded MWIR absorber superpixel with four distinct absorption bands. The pixels absorb energy from a blackbody source and subsequently generate local heat, which is determined by the absorption band of each pixel. Further, the local heat redistributes over the four pixels, leading to a distinct emission spectra, which can be resolved using a microbolometer array. Experiments showed that color-coded absorber pixels are capable for multispectral imaging with detecting temperature ranging from 740–1150 K.

Many tunable metamaterials require cryogenic cooling, and thus may not be feasible for applications requiring compact size or low power usage. However, thermally tunable metamaterials with phase transition materials provide unique access to fascinating possibilities, namely non-volatile switching and fast nano-second switching speed, making them suitable for novel applications. Indeed, metamaterial/superconducting hybrids enable access to high-quality factors that realize broad tunability, and an ultra-high-temperature sensitivity.

5.6   MULTI-TUNABLE METAMATERIAL ABSORBERS

The hybridization of metamaterials with active elements has enabled agile platforms for controlling electromagnetic waves. In addition to modulation schemes using a single external stimulus, it is also possible to take advantage of multiple excitations simultaneously on metamaterials with several active elements, therefore achieving further improvement on the dynamic response. As shown in previous sections, the carrier concentration in semiconductors such as Si can be modulated by optical excitations [70] while the optical conductivity of graphene is changing with an applied bias. For an optically pumped silicon substrate with patterned gold metasurface layer, the generated photo-carriers (electrons or holes) can directly diffuse into a transferred graphene monolayer [82]. A further applied bias between the graphene layer and the high-resistive silicon substrate could significantly shift the Fermi level of the graphene off the Dirac point, increasing the conductivity of graphene in the terahertz range. Using the dual-tuning method, more than 60% of modulation depth on transmission can be obtained with the bias voltage lower than 1 V, which is two orders smaller than that in a single-tuning terahertz metamaterial hybridized with graphene [83]. The graphene layer and silicon layer can also be separately modulated with the electric or optical tuning methods. With a thin spacer layer of strontium titanate (STO) between a patterned graphene metasurface layer and a photo-doped silicon layer, a dynamic terahertz metasurface absorber was recently proposed to achieve 55% of modulation depth on absorption [84]. In addition to the electric and optical modulations, other modulation methods can also be combined together to achieve high-performance tunable absorbers. Through incorporation of a graphene monolayer and temperature dependent media, such as phase change material VO_2 [84, 86], and paraelectric material STO [87], dual-tunable metasurface absorbers undergoing electric and thermal control were also proposed to modulate the multiple properties of absorbers separately. For a broadband absorber consisting of a patterned VO_2 metasurface layer on a graphene monolayer and a dielectric topas spacer, the absorption peak frequency can be adjusted by the voltage applied on graphene layer, whereas the absorbance can be varied with the phase change of VO_2 induced by temperature. Multi-tunable methods could provide more flexibility for control of electromagnetic properties independently and open a new avenue for high-performance metamaterial modulators for imaging, sensing, and communication applications.

5.7 CONCLUSION

In this chapter, we have described the state of the dynamic metamaterial perfect absorbers utilizing various mechanisms including electric, mechanical, thermal, and optical methods. We note that some of these tuning strategies can be applied across a broad portion of the electromagnetic spectrum. Besides, multiple tuning methods can be simultaneously applied on dynamic metamaterials with more freedoms to manipulate the electromagnetic waves. One exception not discussed is the option of magnetic tuning, which is only applicable to some ferromagnetic media operating at microwave frequencies or below [88]. Table 5.1 lists the advantages and limitations of these tuning approaches.

There are significant hurdles which must be overcome in order to make metamaterials feasible for realistic commercial applications. For example, it is still challenging to design and demonstrate a dynamic metamaterial absorber with a full-phase change from 0 to 2π, while the intensity of the outgoing beam is kept the same. In addition, most demonstrated tunable absorbers operate in a relatively narrow spectral range. Increasing the operational spectral range is also of enormous importance for practical applications. Nonetheless, tunable metamaterials have been demonstrated in some cases to achieve state-of-the-art control for manipulation of electromagnetic waves, and thus hold great promise for future technologies. For example, dynamic metamaterials have been used in various technological demonstrations, including single-pixel imaging [9, 45], efficient thermal emission [89], and water desalination [90]. As the metamaterial design paradigm cross pollinates with other state-of-the-art techniques and novel materials and fabricational techniques, such as machine learning, super-resolution imaging, phase change materials, and large-scale nanofabrications, we expect a blossoming of commercial applications utilizing metamaterials across the electromagnetic spectrum.

Table 5.1: Mechanisms for achieving dynamic metamaterial absorbers

Tuning Mechanism	Materials	μW	THz	IR	Vis	Advantages	Limitations
Electrical	Lumped components	✓				High speed, easy implementation, commercial available	Limited scalability to higher frequencies
	Semiconductors	✓	✓	✓	✓	High speed, IC compatible, easy to scale	Limited carrier density change
	Liquid crystals	✓	✓	✓	✓	Large index change, industry compatible	Low speed
Mechanical	Electrostatic deformation		✓	✓	✓	Large modulation index	Low speed, difficult to fabrication
	Bimaterial		✓	✓	✓	Large modulation index, sensitive to temperature	Low speed, material creeping
Optical	Lumped component	✓				Fast, easy implementation, commercial available	Limited scalability to higher frequencies
	Semiconductors	✓	✓	✓	✓	Ultrafast response (ns fs), easy implementation	Difficult for pixelation, bulky system
Thermal	Phase change materials	✓	✓	✓	✓	Large modulation index, easy to demonstrate	Low speed, difficult to control material quality over large scale
	Liquid crystal	✓	✓	✓	✓	Large index change, easy fabrication	Low speed, bulky external heat source

5.8   REFERENCES

[1] Chihhui Wu, Burton Neuner, Gennady Shvets, Jeremy John, Andrew Milder, Byron Zollars, and Steve Savoy. Large-area wide-angle spectrally selective plasmonic absorber. *Physical Review B*, 84(7):075102, August 2011. DOI: 10.1103/physrevb.84.075102 127

[2] Ziqi Miao, Qiong Wu, Xin Li, Qiong He, Kun Ding, Zhenghua An, Yuanbo Zhang, and Lei Zhou. Widely tunable terahertz phase modulation with gate-controlled graphene metasurfaces. *Physical Review X*, 5(4):041027, November 2015. DOI: 10.1103/physrevx.5.041027 127, 133, 134

[3] Junghyun Park, Ju-Hyung Kang, Soo Jin Kim, Xiaoge Liu, and Mark L. Brongersma. Dynamic reflection phase and polarization control in metasurfaces. *Nano Letters*, 17(1):407–413, January 2017. DOI: 10.1021/acs.nanolett.6b04378 127, 132, 133

[4] Xianliang Liu and Willie J. Padilla. Dynamic manipulation of infrared radiation with MEMS metamaterials. *Advanced Optical Materials*, pages n/a–n/a, June 2013. DOI: 10.1002/adom.201300163 127, 137, 138, 139

[5] Kebin Fan, Jonathan Suen, Xueyuan Wu, and Willie J. Padilla. Graphene metamaterial modulator for free-space thermal radiation. *Optics Express*, 24(22):25189, October 2016. DOI: 10.1364/oe.24.025189 127, 134, 135

[6] Mingkai Liu, Mohamad Susli, Dilusha Silva, Gino Putrino, Hemendra Kala, Shuting Fan, Michael Cole, Lorenzo Faraone, Vincent P. Wallace, Willie J. Padilla, David A. Powell, Ilya V. Shadrivov, and Mariusz Martyniuk. Ultrathin tunable terahertz absorber based on MEMS-driven metamaterial. *Microsystems and Nanoengineering*, 3(1), August 2017. DOI: 10.1038/micronano.2017.33 127

[7] Kebin Fan, Jingdi Zhang, Xinyu Liu, Gu-Feng Zhang, Richard D. Averitt, and Willie J. Padilla. Phototunable dielectric Huygens' metasurfaces. *Advanced Materials*, 30(22):1800278, May 2018. DOI: 10.1002/adma.201800278 127

[8] Yong Sun, Wei Tan, Hong-qiang Li, Jensen Li, and Hong Chen. Experimental demonstration of a coherent perfect absorber with pt phase transition. *Physical Review Letters*, 112:143903, April 2014. DOI: 10.1103/physrevlett.112.143903 127

[9] Claire M. Watts, David Shrekenhamer, John Montoya, Guy Lipworth, John Hunt, Timothy Sleasman, Sanjay Krishna, David R. Smith, and Willie J. Padilla. Terahertz compressive imaging with metamaterial spatial light modulators. *Nature Photonics*, 8(8):605–609, June 2014. DOI: 10.1038/nphoton.2014.139 127, 131, 145

[10] Bo Zhu, Zhengbin Wang, Ci Huang, Yijun Feng, Junming Zhao, and Tian Jiang. Polarization insensitive metamaterial absorber with wide incident angle. *Progress in Electromagnetics Research*, 101:231–239, 2010. DOI: 10.2528/pier10011110 127

[11] Bo Zhu, YiJun Feng, Junming Zhao, Ci Huang, Zhengbin Wang, and Tian Jiang. Polarization modulation by tunable electromagnetic metamaterial reflector/absorber. *Optics Express*, 18(22):23196, October 2010. DOI: 10.1364/oe.18.023196 127, 129

[12] Wangren Xu and Sameer Sonkusale. Microwave diode switchable metamaterial reflector/absorber. *Applied Physics Letters*, 103(3):031902, July 2013. DOI: 10.1063/1.4813750 127, 129

[13] Jie Zhao, Qiang Cheng, Jie Chen, Mei Qing Qi, Wei Xiang Jiang, and Tie Jun Cui. A tunable metamaterial absorber using varactor diodes. *New Journal of Physics*, 15(4):043049, April 2013. DOI: 10.1088/1367-2630/15/4/043049 127

[14] Dongju Lee, Heijun Jeong, and Sungjoon Lim. Electronically switchable broadband metamaterial absorber. *Scientific Reports*, 7(1):4891, December 2017. DOI: 10.1038/s41598-017-05330-z 127

[15] Bo Zhu, Yijun Feng, Junming Zhao, Ci Huang, and Tian Jiang. Switchable metamaterial reflector/absorber for different polarized electromagnetic waves. *Applied Physics Letters*, 97(5):051906, August 2010. DOI: 10.1063/1.3477960 128

[16] David Shrekenhamer, Wen-Chen Chen, and Willie J. Padilla. Liquid crystal tunable metamaterial absorber. *Physics Review Letters*, 110(17):177403, April 2013. DOI: 10.1103/physrevlett.110.177403 129, 130

[17] Salvatore Savo, David Shrekenhamer, and Willie J. Padilla. Liquid crystal metamaterial absorber spatial light modulator for THz applications. *Advanced Optical Materials*, 2(3):275–279, March 2014. DOI: 10.1002/adom.201300384 129

[18] Goran Isić, Borislav Vasić, Dimitrios C. Zografopoulos, Romeo Beccherelli, and Radoš Gajić. Electrically tunable critically coupled terahertz metamaterial absorber based on nematic liquid crystals. *Physical Review Applied*, 3(6):064007, June 2015. DOI: 10.1103/physrevapplied.3.064007 129

[19] David Shrekenhamer, John Montoya, Sanjay Krishna, and Willie J. Padilla. Four-color metamaterial absorber THz spatial light modulator. *Advanced Optical Materials*, 1(12):905–909, December 2013. DOI: 10.1002/adom.201300265 131

[20] H. T. Chen, W. J. Padilla, J. M. O. Zide, A. C. Gossard, A. J. Taylor, and R. D. Averitt. Active terahertz metamaterial devices. *Nature*, 444(7119):597–600, 2006. DOI: 10.1038/nature05343 130

[21] Claire M. Watts, Christian C. Nadell, John Montoya, Sanjay Krishna, and Willie J. Padilla. Frequency-division-multiplexed single-pixel imaging with metamaterials. *Optica*, 3(2):133, February 2016. DOI: 10.1364/optica.3.000133 131

[22] Christian C. Nadell, Claire M. Watts, John A. Montoya, Sanjay Krishna, and Willie J. Padilla. Single pixel quadrature imaging with metamaterials. *Advanced Optical Materials*, 4(1):66–69, January 2016. DOI: 10.1002/adom.201500435 131

[23] Jongwon Lee, Seungyong Jung, Pai-Yen Chen, Feng Lu, Frederic Demmerle, Gerhard Boehm, Markus-Christian Amann, Andrea Alù, and Mikhail A. Belkin. Ultrafast electrically tunable polaritonic metasurfaces. *Advanced Optical Materials*, 2(11):1057–1063, November 2014. DOI: 10.1002/adom.201400185 131

[24] S. Law, D. C. Adams, A. M. Taylor, and D. Wasserman. Mid-infrared designer metals. *Optics Express*, 20(11):12155, May 2012. DOI: 10.1364/oe.20.012155 132

[25] Eyal Feigenbaum, Kenneth Diest, and Harry A. Atwater. Unity-order index change in transparent conducting oxides at visible frequencies. *Nano Letters*, 10(6):2111–2116, June 2010. DOI: 10.1021/nl1006307 132

[26] A. Boltasseva and H. A. Atwater. Low-loss plasmonic metamaterials. *Science*, 331(6015):290–291, January 2011. DOI: 10.1126/science.1198258 132

[27] M. A. Noginov, Lei Gu, J. Livenere, G. Zhu, A. K. Pradhan, R. Mundle, M. Bahoura, Yu. A. Barnakov, and V. A. Podolskiy. Transparent conductive oxides: Plasmonic materials for telecom wavelengths. *Applied Physics Letters*, 99(2):021101, July 2011. DOI: 10.1063/1.3604792 132

[28] Gururaj V. Naik, Jongbum Kim, and Alexandra Boltasseva. Oxides and nitrides as alternative plasmonic materials in the optical range [Invited]. *Optical Materials Express*, 1(6):1090, October 2011. DOI: 10.1364/ome.1.001090 132

[29] Fei Yi, Euijae Shim, Alexander Y. Zhu, Hai Zhu, Jason C. Reed, and Ertugrul Cubukcu. Voltage tuning of plasmonic absorbers by indium tin oxide. *Applied Physics Letters*, 102(22):221102, June 2013. DOI: 10.1063/1.4809516 132

[30] Junghyun Park, Ju-Hyung Kang, Xiaoge Liu, and Mark L. Brongersma. Electrically tunable epsilon-near-zero (ENZ) metafilm absorbers. *Scientific Reports*, 5(1):15754, December 2015. DOI: 10.1038/srep15754 132

[31] Jason Horng, Chi-Fan Chen, Baisong Geng, Caglar Girit, Yuanbo Zhang, Zhao Hao, Hans A. Bechtel, Michael Martin, Alex Zettl, Michael F. Crommie, Y. Ron Shen, and Feng Wang. Drude conductivity of Dirac fermions in graphene. *Physical Review B*, 83(16):165113, April 2011. DOI: 10.1103/physrevb.83.165113 132

[32] Aleksei Anopchenko, Long Tao, Catherine Arndt, and Ho Wai Howard Lee. Field-effect tunable and broadband epsilon-near-zero perfect absorbers with deep subwavelength thickness. *ACS Photonics*, 5(7):2631–2637, July 2018. DOI: 10.1021/acsphotonics.7b01373 132

[33] L. Falkovsky and S. Pershoguba. Optical far-infrared properties of a graphene monolayer and multilayer. *Physical Review B*, 76(15):153410, October 2007. DOI: 10.1103/physrevb.76.153410 132

[34] L. A. Falkovsky. Optical properties of graphene. *Journal of Physics: Conference Series*, 129(1):012004, October 2008. DOI: 10.1088/1742-6596/129/1/012004 132

[35] Berardi Sensale-Rodriguez, Rusen Yan, Michelle M. Kelly, Tian Fang, Kristof Tahy, Wan Sik Hwang, Debdeep Jena, Lei Liu, and Huili Grace Xing. Broadband graphene terahertz modulators enabled by intraband transitions. *Nature Communications*, 3(1):780, January 2012. DOI: 10.1038/ncomms1787 132

[36] Long Ju, Baisong Geng, Jason Horng, Caglar Girit, Michael Martin, Zhao Hao, Hans A. Bechtel, Xiaogan Liang, Alex Zettl, Y. Ron Shen, and Feng Wang. Graphene plasmonics for tunable terahertz metamaterials. *Nature Nanotechnology*, 6(10):630–4, October 2011. DOI: 10.1038/nnano.2011.146 132

[37] Jianfa Zhang, Chucai Guo, Ken Liu, Zhihong Zhu, Weimin Ye, Xiaodong Yuan, and Shiqiao Qin. Coherent perfect absorption and transparency in a nanostructured graphene film. *Optics Express*, 22(10):12524, May 2014. DOI: 10.1364/oe.22.012524 133

[38] Yu Tong Zhao, Bian Wu, Bei Ju Huang, and Qiang Cheng. Switchable broadband terahertz absorber/reflector enabled by hybrid graphene-gold metasurface. *Optics Express*, 25(7):7161, April 2017. DOI: 10.1364/oe.25.007161 133, 134

[39] Yu Yao, Raji Shankar, Mikhail A. Kats, Yi Song, Jing Kong, Marko Loncar, and Federico Capasso. Electrically tunable metasurface perfect absorbers for ultrathin mid-infrared optical modulators. *Nano Letters*, 14(11):6526–32, November 2014. DOI: 10.1021/nl503104n 134, 135

[40] Borislav Vasić and Radoš Gajić. Graphene induced spectral tuning of metamaterial absorbers at mid-infrared frequencies. *Applied Physics Letters*, 103(26):261111, December 2013. DOI: 10.1063/1.4858459 134

[41] Gang Yao, Furi Ling, Jin Yue, Chunya Luo, Jie Ji, and Jianquan Yao. Dual-band tunable perfect metamaterial absorber in the THz range. *Optics Express*, 24(2):1518, January 2016. DOI: 10.1364/oe.24.001518 134

[42] Hongju Li, Lingling Wang, and Xiang Zhai. Tunable graphene-based mid-infrared plasmonic wide-angle narrowband perfect absorber. *Scientific Reports*, 6(1):36651, December 2016. DOI: 10.1038/srep36651 134

[43] Lujun Huang, Guoqing Li, Alper Gurarslan, Yiling Yu, Ronny Kirste, Wei Guo, Junjie Zhao, Ramon Collazo, Zlatko Sitar, Gregory N. Parsons, Michael Kudenov, and Linyou Cao. Atomically thin MoS$_2$ narrowband and broadband light superabsorbers. *ACS Nano*, 10(8):7493–7499, August 2016. DOI: 10.1021/acsnano.6b02195 134

[44] Michelle C. Sherrott, Philip W. C. Hon, Katherine T. Fountaine, Juan C. Garcia, Samuel M. Ponti, Victor W. Brar, Luke A. Sweatlock, and Harry A. Atwater. Experimental demonstration of >230° phase modulation in gate-tunable graphene-gold recon-

figurable mid-infrared metasurfaces. *Nano Letters*, 17(5):3027–3034, May 2017. DOI: 10.1021/acs.nanolett.7b00359 134

[45] Kebin Fan, Jonathan Y. Suen, and Willie J. Padilla. Graphene metamaterial spatial light modulator for infrared single pixel imaging. *Optics Express*, 25(21):25318, October 2017. DOI: 10.1364/oe.25.025318 134, 145

[46] Beibei Zeng, Zhiqin Huang, Akhilesh Singh, Yu Yao, Abul K. Azad, Aditya D. Mohite, Antoinette J. Taylor, David R. Smith, and Hou-Tong Chen. Hybrid graphene metasurfaces for high-speed mid-infrared light modulation and single-pixel imaging. *Light: Science and Applications*, 7(1):51, December 2018. DOI: 10.1038/s41377-018-0055-4 134, 135

[47] Andrei Andryieuski and Andrei V. Lavrinenko. Graphene metamaterials based tunable terahertz absorber: Effective surface conductivity approach. *Optics Express*, 21(7):9144, April 2013. DOI: 10.1364/oe.21.009144 134

[48] Yu Yao, Mikhail A. Kats, Raji Shankar, Yi Song, Jing Kong, Marko Loncar, and Federico Capasso. Wide wavelength tuning of optical antennas on graphene with nanosecond response time. *Nano Letters*, 14(1):214–219, December 2013. DOI: 10.1021/nl403751p 135

[49] David A. Powell, Mikhail Lapine, Maxim V. Gorkunov, Ilya V. Shadrivov, and Yuri S. Kivshar. Metamaterial tuning by manipulation of near-field interaction. *Physical Review B*, 82:155128, October 2010. DOI: 10.1103/physrevb.82.155128 136

[50] Mikhail Lapine, Ilya V. Shadrivov, David A. Powell, and Yuri S. Kivshar. Magnetoelastic metamaterials. *Nature Materials*, 11(1):30–3, January 2012. DOI: 10.1038/nmat3168 136

[51] W. M. Zhu, A. Q. Liu, X. M. Zhang, D. P. Tsai, T. Bourouina, J. H. Teng, X. H. Zhang, H. C. Guo, H. Tanoto, T. Mei, G. Q. Lo, and D. L. Kwong. Switchable magnetic metamaterials using micromachining processes. *Advanced Materials*, 23(15):1792–+, 2011. DOI: 10.1002/adma.201004341 136

[52] H. Tao, A. C. Strikwerda, K. Fan, W. J. Padilla, X. Zhang, and R. D. Averitt. Reconfigurable terahertz metamaterials. *Physics Review Letters*, 103(14), 2009. DOI: 10.1103/physrevlett.103.147401 136

[53] Imogen M. Pryce, Koray Aydin, Yousif A. Kelaita, Ryan M. Briggs, and Harry A. Atwater. Highly strained compliant optical metamaterials with large frequency tunability. *Nano Letters*, 10(10):4222–7, October 2010. DOI: 10.1021/nl102684x 136

[54] E. Ekmekci, A. C. Strikwerda, K. Fan, G. Keiser, X. Zhang, G. Turhan-Sayan, and R. D. Averitt. Frequency tunable terahertz metamaterials using broadside coupled split-ring resonators. *Physical Review B*, 83(19):193103, May 2011. DOI: 10.1103/physrevb.83.193103 136

[55] Jining Li, Charan M. Shah, Withawat Withayachumnankul, Benjamin S.-Y. Ung, Arnan Mitchell, Sharath Sriram, Madhu Bhaskaran, Shengjiang Chang, and Derek Abbott. Mechanically tunable terahertz metamaterials. *Applied Physics Letters*, 102(12):121101, March 2013. DOI: 10.1063/1.4773238 136

[56] Fabio Alves, Dragoslav Grbovic, Brian Kearney, and Gamani Karunasiri. Microelectromechanical systems bimaterial terahertz sensor with integrated metamaterial absorber. *Optics Letters*, 37(11):1886, June 2012. DOI: 10.1364/ol.37.001886 136, 137

[57] Fabio Alves, Dragoslav Grbovic, Brian Kearney, Nickolay V. Lavrik, and Gamani Karunasiri. Bi-material terahertz sensors using metamaterial structures. *Optics Express*, 21(11):13256, May 2013. DOI: 10.1364/oe.21.013256 136

[58] Xinyu Liu and Willie J. Padilla. Reconfigurable room temperature metamaterial infrared emitter. *Optica*, 4(4):430, April 2017. DOI: 10.1364/optica.4.000430 136, 138, 139

[59] Weiren Zhu, Fajun Xiao, Ming Kang, and Malin Premaratne. Coherent perfect absorption in an all-dielectric metasurface. *Applied Physics Letters*, 108(12):121901, March 2016. DOI: 10.1063/1.4944635 136

[60] P. G. Datskos, N. V. Lavrik, and S. Rajic. Performance of uncooled microcantilever thermal detectors. *Review of Scientific Instruments*, 75(4):1134–1148, 2004. DOI: 10.1063/1.1667257 136, 142

[61] Prakash Pitchappa, Chong Pei Ho, Piotr Kropelnicki, Navab Singh, Dim-Lee Kwong, and Chengkuo Lee. Micro-electro-mechanically switchable near infrared complementary metamaterial absorber. *Applied Physics Letters*, 104(20):201114, May 2014. DOI: 10.1063/1.4879284 137

[62] W. J. Padilla, A. J. Taylor, and R. D. Averitt. Dynamical electric and magnetic metamaterial response at terahertz frequencies. *Physics Review Letters*, 96(10):107401, March 2006. DOI: 10.1103/physrevlett.96.107401 139

[63] Hou-Tong Chen, Willie J. Padilla, Joshua M. Zide, Seth R. Bank, Arthur C. Gossard, Antoinette J. Taylor, and Richard D. Averitt. Ultrafast optical switching of terahertz metamaterials fabricated on ErAs/GaAs nanoisland superlattices. *Optics Letters*, 32(12):1620, June 2007. DOI: 10.1364/ol.32.001620 139

[64] H. T. Chen, J. F. O'Hara, A. K. Azad, A. J. Taylor, R. D. Averitt, D. B. Shrekenhamer, and W. J. Padilla. Experimental demonstration of frequency-agile terahertz metamaterials. *Nature Photonics*, 2(5):295–298, 2008. DOI: 10.1038/nphoton.2008.52 139

[65] Kebin Fan, Andrew C. Strikwerda, Xin Zhang, and Richard D. Averitt. Three-dimensional broadband tunable terahertz metamaterials. *Physical Review B*, 87(16):161104, April 2013. DOI: 10.1103/physrevb.87.161104 139

[66] Kebin Fan, Xiaoguang Zhao, Jingdi Zhang, Kun Geng, George R. Keiser, Huseyin R. Seren, Grace D. Metcalfe, Michael Wraback, Xin Zhang, and Richard D. Averitt. Optically tunable terahertz metamaterials on highly flexible substrates. *IEEE Transactions on Terahertz Science and Technology*, 3(6):702–708, November 2013. DOI: 10.1109/tthz.2013.2285619 139

[67] David J. Cho, Wei Wu, Ekaterina Ponizovskaya, Pratik Chaturvedi, Alexander M. Bratkovsky, Shih-Yuan Wang, Xiang Zhang, Feng Wang, and Y. Ron Shen. Ultrafast modulation of optical metamaterials. *Optics Express*, 17(20):17652, September 2009. DOI: 10.1364/oe.17.017652 140

[68] Huseyin R. Seren, George R. Keiser, Lingyue Cao, Jingdi Zhang, Andrew C. Strikwerda, Kebin Fan, Grace D. Metcalfe, Michael Wraback, Xin Zhang, and Richard D. Averitt. Optically modulated multiband terahertz perfect absorber. *Advanced Optical Materials*, 2(12):1221–1226, December 2014. DOI: 10.1002/adom.201400197 140, 141

[69] Jun Chen, Jack Ng, Zhifang Lin, and C. T. Chan. Optical pulling force. *Nature Photonics*, 5(9):531–534, 2011. DOI: 10.1038/nphoton.2011.153 140

[70] Xiaoguang Zhao, Kebin Fan, Jingdi Zhang, Huseyin R. Seren, Grace D. Metcalfe, Michael Wraback, Richard D. Averitt, and Xin Zhang. Optically tunable metamaterial perfect absorber on highly flexible substrate. *Sensors and Actuators A: Physical*, 231:74–80, July 2015. DOI: 10.1016/j.sna.2015.02.040 140, 144

[71] A. Zylbersztejn and N. F. Mott. Metal-insulator transition in vanadium dioxide. *Physical Review B*, 11:4383–4395, June 1975. DOI: 10.1103/physrevb.11.4383 142

[72] Chen Ang, A. S. Bhalla, and L. E. Cross. Dielectric behavior of paraelectric $ktao_3$, $catio_3$, and $(Ln_{1/2}na_{1/2})tio_3$ under a dc electric field. *Physical Review B*, 64:184104, October 2001. DOI: 10.1103/PhysRevB.64.184104 142

[73] R. W. Cunningham and L. E. Cross. Intrinsic concentration and heavy-hole mass in InSb. *Journal of Applied Physics*, 41(4):1804, March 1970. DOI: 10.1063/1.1659107 142

[74] Jaronie Mohd Jani, Martin Leary, Aleksandar Subic, and Mark A. Gibson. A review of shape memory alloy research, applications, and opportunities. *Materials and Design*, 56:1078–1113, April 2014. DOI: 10.1016/j.matdes.2013.11.084 142

[75] T. Driscoll, S. Palit, M. M. Qazilbash, M. Brehm, F. Keilmann, B. G. Chae, S. J. Yun, H. T. Kim, S. Y. Cho, N. M. Jokerst, D. R. Smith, and D. N. Basov. Dynamic tuning of an infrared hybrid-metamaterial resonance using vanadium dioxide. *Applied Physics Letters*, 93(2), 2008. DOI: 10.1063/1.2956675 142

[76] Mengkun Liu, Harold Y. Hwang, Hu Tao, Andrew C. Strikwerda, Kebin Fan, George R. Keiser, Aaron J. Sternbach, Kevin G. West, Salinporn Kittiwatanakul, Jiwei Lu, Stuart A. Wolf, Fiorenzo G. Omenetto, Xin Zhang, Keith A. Nelson, and Richard D. Averitt. Terahertz-field-induced insulator-to-metal transition in vanadium dioxide metamaterial. *Nature*, 487(7407):345–8, July 2012. DOI: 10.1038/nature11231 142

[77] Qi-Ye Wen, Huai-Wu Zhang, Qing-Hui Yang, Zhi Chen, Yang Long, Yu-Lan Jing, Yuan Lin, and Pei-Xin Zhang. A tunable hybrid metamaterial absorber based on vanadium oxide films. *Journal of Physics D: Applied Physics*, 45(23):235106, June 2012. DOI: 10.1088/0022-3727/45/23/235106 142

[78] Mikhail A. Kats, Deepika Sharma, Jiao Lin, Patrice Genevet, Romain Blanchard, Zheng Yang, M. Mumtaz Qazilbash, D. N. Basov, Shriram Ramanathan, and Federico Capasso. Ultra-thin perfect absorber employing a tunable phase change material. *Applied Physics Letters*, 101(22):221101, November 2012. DOI: 10.1063/1.4767646 142

[79] Hao Wang, Yue Yang, and Liping Wang. Switchable wavelength-selective and diffuse metamaterial absorber/emitter with a phase transition spacer layer. *Applied Physics Letters*, 105(7):071907, August 2014. DOI: 10.1063/1.4893616 142

[80] Andreas Tittl, Ann-Katrin U. Michel, Martin Schäferling, Xinghui Yin, Behrad Gholipour, Long Cui, Matthias Wuttig, Thomas Taubner, Frank Neubrech, and Harald Giessen. A switchable mid-infrared plasmonic perfect absorber with multispectral thermal imaging capability. *Advanced Materials*, 27(31):4597–4603, August 2015. DOI: 10.1002/adma.201502023 142, 143

[81] Kostiantyn Shportko, Stephan Kremers, Michael Woda, Dominic Lencer, John Robertson, and Matthias Wuttig. Resonant bonding in crystalline phase-change materials. *Nature Materials*, 7(8):653–658, August 2008. DOI: 10.1038/nmat2226 142

[82] Quan Li, Zhen Tian, Xueqian Zhang, Ningning Xu, Ranjan Singh, Jianqiang Gu, Peng Lv, Lin-Bao Luo, Shuang Zhang, Jiaguang Han, and Weili Zhang. Dual control of active graphene–silicon hybrid metamaterial devices. *Carbon*, 90:146–153, August 2015. DOI: 10.1016/j.carbon.2015.04.015 144

[83] Seung Hoon Lee, Muhan Choi, Teun-Teun Kim, Seungwoo Lee, Ming Liu, Xiaobo Yin, Hong Kyw Choi, Seung S. Lee, Choon-Gi Choi, Sung-Yool Choi, Xiang Zhang, and Bumki Min. Switching terahertz waves with gate-controlled active graphene metamaterials. *Nature Materials*, 11(11):936–941, September 2012. DOI: 10.1038/nmat3433 144

[84] Jinfeng Wang, Tingting Lang, Zhi Hong, Tingting Shen, and Gangqi Wang. Tunable terahertz metamaterial absorber based on electricity and light modulation modes. *Optical Materials Express*, 10(9):2262, August 2020. DOI: 10.1364/ome.402541 144

[85] Tongling Wang, Yuping Zhang, Huiyun Zhang, and Maoyong Cao. Dual-controlled switchable broadband terahertz absorber based on a graphene-vanadium dioxide metamaterial. *Optical Materials Express*, 10(2):369, January 2020. DOI: 10.1364/ome.383008

[86] Hui Li and Jiang Yu. Active dual-tunable broadband absorber based on a hybrid graphene-vanadium dioxide metamaterial. *OSA Continuum*, 3(8):2143, August 2020. DOI: 10.1364/osac.397243 144

[87] Renxia Ning, Zhiqiang Xiao, Zhenhai Chen, and Wei Huang. Dual-tunable polarization insensitive electromagnetically induced transparency in metamaterials. *Journal of Electronic Materials*, 50(7):3916–3922, April 2021. DOI: 10.1007/s11664-020-08692-9 144

[88] Yongjun Huang, GuangJun Wen, Weiren Zhu, Jian Li, Li-Ming Si, and Malin Pre-Maratne. Experimental demonstration of a magnetically tunable ferrite based metamaterial absorber. *Optics Express*, 22(13):16408, June 2014. DOI: 10.1364/oe.22.016408 145

[89] Chun-Chieh Chang, Wilton J. M. Kort-Kamp, John Nogan, Ting S. Luk, Abul K. Azad, Antoinette J. Taylor, Diego A. R. Dalvit, Milan Sykora, and Hou-Tong Chen. High-temperature refractory metasurfaces for solar thermophotovoltaic energy harvesting. *Nano Letters*, 18(12):7665–7673, December 2018. DOI: 10.1021/acs.nanolett.8b03322 145

[90] Lin Zhou, Yingling Tan, Jingyang Wang, Weichao Xu, Ye Yuan, Wenshan Cai, Shining Zhu, and Jia Zhu. 3D self-assembly of aluminium nanoparticles for plasmon-enhanced solar desalination. *Nature Photonics*, 10(6):393–398, April 2016. DOI: 10.1038/npho-ton.2016.75 145

CHAPTER 6

Applications

A major motivating force for the use of electromagnetic absorbers occurred in World War II, when it was desired to conceal both ships and submarines from microwave radar detection. However, use of absorbers for other applications and in order bands of the electromagnetic spectrum did not occur until relatively recently, primarily due to the maturation of both nano-fabricational techniques, and commercialization of computational electromagnetic (CEM) software. In previous chapters, we have shown that the development of metamaterial absorbers has enabled extraordinary control of absorptivity or emissivity spectra over much of the electromagnetic spectrum from the microwave range through infrared to the ultraviolet wavelengths, thus suggesting their use for novel applications. In this chapter, we will review a few examples of metamaterial absorbers used for potential applications in imaging, sensing, solar energy harvesting, and color engineering.

6.1 BIOCHEMICAL SENSING

For a material located in a resonating cavity system, perturbation theory indicates that a change in its permittivity leads to a modification of the resonant frequency [1, 2]. The sub-wavelength nature of metamaterials causes both high field localization and field enhancement. Thus, locating materials within field enhancing regions of metamaterials, sensitivity to the change of refractive index may be further enhanced. Additionally, according to temporal-coupled mode theory, a slight change of the material loss in a perfect absorber could breach the critical coupling condition, thereby significantly altering the scattering response, both in amplitude and phase [3].

Since the early development of metamaterial absorbers, they have been used to fashion structures for sensing applications, e.g., glucose sensing [4] and Hydrogen sensing [7]. The first experimental demonstration of chemical sensing was performed with plasmonic perfect absorber operating in the near-infrared region [4]. As shown in Fig. 6.1a, the infrared perfect absorber consists of arrayed 20-nm-thick gold disks with a diameter of 352 nm residing on a 30-nm-thick MgF_2 spacer, which is supported by a gold mirror. Spectroscopic characterization in air at infrared wavelength showed a polarization-independent absorbance at a wavelength of 1.6 μm as high as 99%, corresponding to a near-zero reflectance. With water ($n = 1.312$) applied to the sample surface, an observable change of reflectance from 1–28.7% was observed at 185.6 THz,

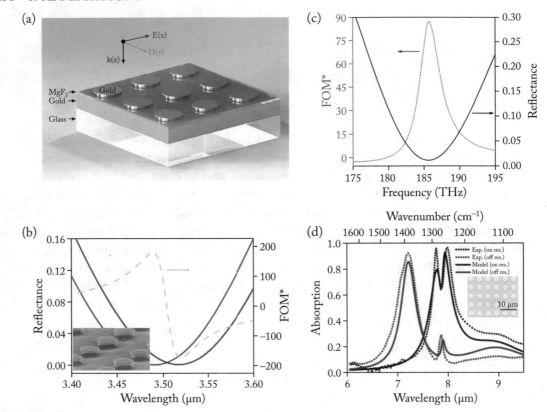

Figure 6.1: (a) Schematic of an infrared perfect absorber for sensing application. It consists of a gold ground plane, MgF_2 spacer, and gold disk array on top of a glass substrate. (b) Experimental FOM* (dashed curve) and reflectance (solid curve) as a function of frequency. (c) Calculated FOM* of an infrared perfect absorber as the refractive index of glucose solution changes from 1.312 (blue) to 1.322 (red). The inset SEM image shows an oblique view of an array of mushroom-capped plasmonic absorber. (d) Experimental (dotted line) and analytically modeled (solid line) absorption spectra of perfect absorbers for geometric resonances coincident (3.61 μm) and spectrally removed (2.68 μm) from the material resonance in the spin-on glass. The inset shows an SEM image of the fabricated infrared perfect absorber with a spin-on-glass as the spacer. Reprinted with permission from: (a) and (b) Ref. [4] © 2010 ACS; (c) Ref. [5] © 2015 Wiley-VCH; (d) Ref. [6] © 2012 IEEE.

due to a change in the index of refraction. The figure of merit (FOM*) is defined as

$$FOM^* = \text{Max} \left| \frac{dI(\lambda)/dn(\lambda)}{I(\lambda)} \right|, \tag{6.1}$$

where $dI(\lambda)/dn(\lambda)$ is the relative intensity change at a fixed wavelength induced by a refractive-index change, and λ_0 is chosen at maximum FOM*, reaches a value of 87 for water with index change of 0.312 as shown in Fig. 6.1b. In terms of resonant frequency shift, the sensor exhibits about 400-nm per refractive index unit (RIU). In addition to sensing with conventional plasmonic metamaterial structures, it was also shown that Babinet plasmonic absorbers could further improve refractive index sensing performance by approximately 50% [8]. Such an absorber design can be scaled to other frequency ranges, such as terahertz, for sensing chemicals such as photoresist [9].

In most designs, however, only the surface of the planar metamaterial is accessible to chemicals for sensing, whereas the field confined regions are restricted to the horizontal gaps of the plasmonic structures, such as the field enhancement in the spacer region. Bhattarai and colleagues demonstrated a mushroom-capped plasmonic perfect absorber, in which the dielectric spacer is partially etched away, thus enabling permeation of chemicals into the spacer region [5]. The inset of Fig. 6.1c shows an SEM image of the fabricated mushroom-capped infrared absorber with each gold disk supported by a polymer cylinder, which was laterally etched in an O_2 plasma. Due to the significant contact area of the chemical with the plasmonic structure, simulations showed that the sensitivity, at 2.71 μm, after applying water onto the sample could reach 2513-nm/RIU, which is about 5 times larger than the first generation of metamaterial absorber sensor [4]. Further, the FOM* as another measure of the sensor performance, is also significantly improved to 179; see Fig. 6.1c.

It should be mentioned that chemical sensors based on resonance shifts do not guarantee the identification of chemicals from others which posses a similar refractive index. To further enhance the specificity of sensors, Mason and his colleagues designed a perfect absorber with the resonance overlapping with the characteristic vibrational modes of molecules [6]. Due to surface enhancement from excited plasmons, the measured chemical absorption can be ten times stronger than that with a coating on a metal film [10]. As shown in Fig. 6.1d, the strong coupling between these two resonances reveals an observable anti-crossing with an energy splitting due to the hybridization of the optical and vibrational modes.

6.2   ENERGY HARVESTING

6.2.1   PHOTODETECTION VIA HOT ELECTRON HARVESTING

The manipulation of excited coherently oscillating electrons in metal, i.e., surface plasmons (SPs), have enabled a myriad of techniques and applications in manipulating light at a deep-subwavelength scale, such as modulators, plasmonic enhanced photovoltaic devices, superresolution imaging, and photodetectors. Recent progress has shown that nonradiatively decaying electrons can be exploited for photodetection, in which the excited hot electrons hop through Schottky barriers formed at the metal-semiconductor interface, which are then injected into the conduction band of semiconductors. Therefore, the bandwidth of photodetection is only limited by the Schottky barrier height rather than the bandgap of the semiconductor. Utilizing a plasmonic antenna fabricated on silicon substrate, Knight et al. achieved photodetection in the telecommunication regime, the photon energy of which is well below the bandgap of the semiconductor [11]. The measured photocurrent exhibits a direct correlation to the absorption spectrum of antennas.

However, the bandwidth of SP-based resonators is intrinsically narrow, thus rendering a limited responsivity. Lin and colleagues devised a broadband deep trench/thin metal (DTTM) absorber composed of etched silicon trenches covered by a thin layer of gold on the surface, as shown in Fig. 6.2a [12]. Such an architecture is free of a metal ground plane, and circumvents the challenge of growing a high-quality semiconductor on top of a metal film. Still, the deep-trench structure maintains high absorptivity over 80% from about 1375–1475-nm from generating surface plasmon polariton (SPP) waves which are confined in the trench cavity. Compared to nanoantennas and dot/hole arrays, the peak absorptivity is about four times and one order of magnitude higher, respectively. The peak responsivity for optimized structures is measured close to 3.8 mA/W, which is about 380 times larger than that obtained by Knight from nanorods. In the same year, Li and Valentine also demonstrated an inverse structure to Lin's device [13]. As shown in Fig. 6.2b, the broadband MPA is composed of etched silicon stripes or patches with a 15-nm gold thin film covering over the surface and sidewalls. The Fabry-Pérot resonance is formed from the upper and lower plasmonic resonators, and results in a broadband absorption spectrum over 80% from 1250–1500 nm. The efficiency of the hot electron hopping process is significantly enhanced, and is attributed to the ultrathin 15-nm metal layer. The measured photoresponsivity is larger than 1.8 mA from 1200–1500 nm, as shown in Fig. 6.2d.

Despite the achievement of large photoresponsivity from periodically patterned meta-material absorbers, future commercial implementation of these nanostructures will rely on the advances of the nano-lithography techniques to further lower the cost, and to accelerate the fabrication process for large-area patterns. With controlled evaporation process or thermal annealing processes on ultrathin gold films, randomly distributed irregular gold nanostructures can be synthesized directly, enabling broadband absorption [14–16]. Lu and co-workers experimentally demonstrated broadband hot-electron photodetection with a collection of randomly dispersed gold nanoparticles formed on gold film supported TiO_2 spacer—a wide-bandgap semiconductor [14]. The surface coverage and size dispersion of the nanoparticles was achieved through control of the initial Au film thickness, and the thermal annealing conditions. After the thermal treatment in air at 400°C for 3 hours, the 5-nm-thick pre-deposited Au film on a 50-nm-thick TiO_2 spacer formed nanoparticles dispersed over about 35% of the surface area with a Gaussian distribution of size centered at 12-nm as shown in Fig. 6.2e–f. Measured spectra exhibits remarkably high absorption over 90% in the entire visible spectrum from 400–750-nm (Fig. 6.2g). In addition, the optical absorption of the optimized structure shows trivial polarization and angular dependency. The photocurrent density under the visible illumination was measured as high as 15.8 $\mu A/cm^2$. Using a similar concept, Wen et al. demonstrated a broadband absorber consisting of ultrathin plasmonic metal coated disordered silicon nanohole

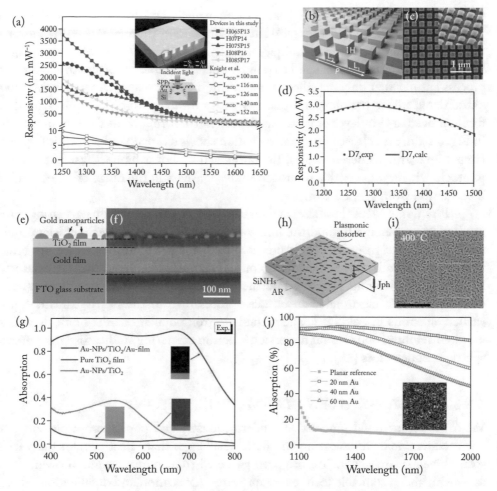

Figure 6.2: (a) Measured photoresponsivity of the DTTM absorber under bias of 0 V and comparison to Knight's work. The inset illustrates the DTTM device configuration. (b) Schematic of a broadband absorber with two different silicon patches in an array. $L1 = 185$-nm, $L2 = 225$-nm, $P = 680$-nm, $H = 160$-nm. (c) SEM image of the fabricated device. (d) Comparison between calculated (line) and measured (circles) photoresponsivity spectra of the broadband MPA. (e, f) Schematic and SEM image of the broadband absorber with randomly dispersed gold nanoparticles on a thin TiO_2 film. (g) Measured absorption spectra of samples with different coatings. The inset shows the photographic images of the prepared samples. (h) Schematic of a hot-electron photodetector with disordered nanoholes. (i) SEM image of the fabricated device using Au templates dewetted at 400°C. (j) Measured absorption spectra with different thickness of Au coating. The inset shows simulated electric field distribution in the region shown in dashed green box in (i). Reprinted with permission from: (a) Ref. [12] © 2014 NPG; (b)–(d) Ref. [13] © 2010 ACS; (e)–(g) Ref. [14] © 2016 NPG; (h) and (i) Ref. [15] © 2018 ACS.

structures, as shown in Fig. 6.2h–i, supporting many hot spots with enhanced electric field [15]. The disordered nanoholes were fabricated mainly using thermal dewetting and metal assisted chemical etching (MACE) techniques. In the rapid thermal dewetting process, a continuous 5-nm gold thin film was changed to different nanostructured morphologies. Then, the MACE process ensures that only the silicon underneath the Au catalyst film was etched away, resulting a disordered silicon nanohole morphology with a mean depth of 100 nm. After the Au catalyst film removed, a thin layer of gold was deposited conformally on the random holes. As shown in Fig. 6.2j, the spectroscopic characterization shows a broadband absorption greater than 80% from 1.1–2 μm. Under zero bias, the optimized design exhibits a fast hot electron mediated photoelectric response with a photocurrent responsivity in the range of 1.5–13 mA/W with the illumination from 1.1–1.5 μm.

Plasmonic MPAs have shown their capability to enable photon harvesting through photon to electron conversion beyond the limit of semiconductors. To create efficient photodetectors, the judicious design of MPAs ensures two fundamental requisites which are broadband high absorption for collecting sufficient photons and large field enhancement to accelerate plasmons into energetic electrons injecting into the conduction band of semiconductors. The excitation of hot plasmonic electrons is an ultrafast process due to the nearly ballistic transport of the excited electrons in gold [17]. The excited electron populations then quickly decay over a few hundred femtoseconds through electron-electron scattering, indicating the potential as high-speed photodetectors [18].

6.2.2 SOLAR-DRIVEN STEAM GENERATION

With the recent rapid progress of modern industry and population explosion, water scarcity has become one of the most urgent global challenges that affect people's daily lives. Recently, solar-driven steam generation has been garnered tremendous attention owing to its potential as a green and sustainable technology for water purification and desalination. There are several factors that impact the performance of solar generation, including water transport and thermal management. However, the broadband absorption sunlight is singularly critical, and possesses great potential to improve the efficiency [19].

Recent plasmonic advances have demonstrated devices with significant absorption enhancement and heat localization, which can be beneficial to solar steam generation. However, due to the relatively small Drude damping factor of noble metal nanoparticles, the absorption bandwidth of spherical or cubic-like nanoparticles is limited. To overcome such constraints, theoretical analysis showed that some nanoparticles, such as TiN [22, 23], indium [24], support lossy plasmonic resonances in the visible range, and exhibit better broadband solar absorption than gold nanoparticles. One the other hand, through judicious engineering of plasmonic nanostructures with more dispersive dimensions, ultra-broadband plasmonic absorbers can be achieved to boost the solar steaming efficiency. Bae and co-workers demonstrated an ultra-broadband absorber with multi-scale metallic funnel structures on a flexible membrane [20].

Via controlled wet etching of an anodic aluminium oxide (AAO) template, the etched alumina nanowires collapsed together forming self-aggregated nanowire bundles due to the water-air surface tension, as shown in Fig. 6.3a–b. After metal sputtering, the funnelled structure with varied nanogaps among the nanowires can adiabatically couple light into surface plasmons, thus producing high absorption up to 17 μm (Fig. 6.3c). Experiments showed efficient steam generation with a solar thermal conversion efficiency up to 57% at 20 kW/m^2. In later studies, as shown in Fig. 6.3d–g, Zhou et al. demonstrated that an ultra-broadband absorber could be obtained without delicate etching of AAO templates, but directly evaporating metallic nanoparticles into the nano-pores [21, 25, 26]. Due to the hybridization of the localized surface plasmonic resonances and non-radiative plasmon decay, the self-assembled nanostructures in the porous membrane achieved absorptivity of 99% across the wavelength from 400 nm to 10 μm, as shown in Fig. 6.3h. As a result of the efficient and broadband light absorption, strongly localized heating, and porous structures, which allow continuous stream flows by the generated heat, such a self-assembled broadband plasmonic absorber achieved the efficiency of solar steam generation and solar desalination over 90% with 4-sun irradiation.

Given the brittle AAO template, relatively large cost, and potential scalability challenges, other alternative strategies with decorating gold nanoparticles onto other mesoporous and hierarchical structures, such as carbonized biomass [27], diverse carbon materials [28, 30, 31], have been investigated. The lower thermal conductivity of these materials can further localized the generated heat, thus leading to higher solar energy conversion efficiency. The closely packed nanochannels in AAO templates can be blocked after a long exposure [32], whereas a matrix with numerous aligned micro- and nanochannels can effectively conduct water from the bottom to the evaporation surface from the capillary effect. These methods have enabled a range of potential applications to tackle the global challenge of water crisis.

6.2.3　SELECTIVE EMITTER FOR THERMOPHOTOVOLTAICS (TPV)

Salvaging energy from radiated heat ($> 1000°C$) is known to be more efficient than conversion of energy from nature resources, such as fossil fuels and coal where over 50% of the total produced power lost [33]. With the aid of thermophotovoltaic technology, radiated heat with temperature over 1000°C which would normally be lost to the environment, can be potentially collected and harvested. Through implementation of this technique, the radiated thermal energy will be concentrated and converted to heat inside of a structure which possesses a spectral emission tailored to match the bandgap of a specific photovoltaic cell—example include GaSb, InGaAs, and InGaAsSb. However, such spectrally selective absorbers or emitters are difficult to obtain directly from nature materials. Over the past decades, extensive studies have been performed toward tailorable emitters for TPVs using photonic crystals [34] and planar Fabry–Perot based structures [35]. Electromagnetic metamaterials have shown their power to engineer the emission spectrum through adjusting the geometry and space filling properties of metamaterial unit cells [4, 33, 36–46].

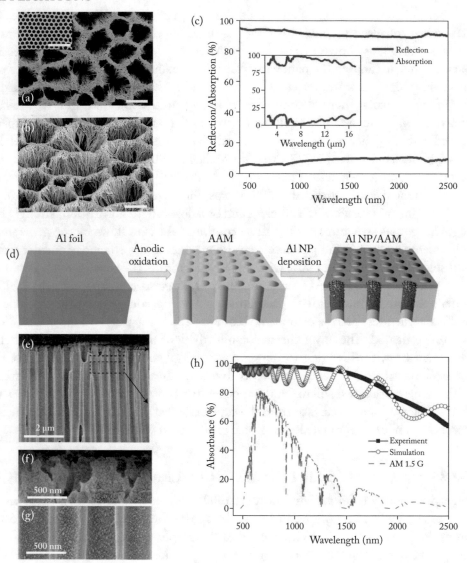

Figure 6.3: SEM images of the top-view, (a) and a bird's-eye view (b) of the black gold membrane. The inset shows the ordered hexagonal array of the AAO template. Scale bars, 2 μm. (c) Measured optical response of the a black gold membrane peeled off with aluminum tape. The inset shows represents reflection and absorption from 2.5–17 μm. (d) Fabrication process of the broadband absorber from an aluminum foil. (e)–(g) SEM images of cross-sectional view of the fabricated sample. (h) Measured (black dots) and simulated (red dots) absorption of the aluminum-based plasmonic absorber. The blue dashed curve is the normalized spectral solar irradiance density of air mass 1.5 global (AM1.5 G) tilted solar spectrum. Reprinted with permission from: (a)–(c) Ref. [20] © 2015 NPG; (d)–(h) Ref. [21] © 2016 NPG.

A TPV system usually consists of a hot emitter with spectrally selective emission, paired with a low-bandgap semiconductor, which is used to convert the emitted photons into electrons. The efficiency of the TPV cell is governed by the Shockley–Queisser limit, which is governed by the difference between the emitted photon energy and the band gap of the semiconductor. To increase TPV efficiency, Liu et al. proposed a metamaterial absorber supercell, of which the emittance spectrum is designed to follow the wavelength dependent external quantum efficiency (EQE) of GaSb with photon energy larger than the band gap [47]. In the designed structure, 4 different metallic cross resonators were arranged into a 2×2 unit-cell, which were then used to form a 2×2 supercell with 16 total elements. As shown in Fig. 3.9c, the emittance spectrum is fashioned to match the EQE of GaSb as close as possible. The emittance spectrum can be independently controlled by the geometry of sublattice cell, which demonstrates a path to manipulate the emission spectra in an arbitrary fashion.

In addition to the challenges of increasing TPV efficiency through emittance band engineering, another area ripe for efficiency gains involves the operational temperature of the emitter, which requires the emitter hot enough to emit a significant amount of photons above the bandgap of the semiconductor. As a result, materials used for the perfect absorbers should be able to withstand high temperature operation for long duration without degradation. Noble metals, such as gold, copper and silver, are not practical for such purposes, due to their low melting points. Instead, refractory metals, such as tungsten (W) [40, 43, 46, 49], platinum [39, 45], titanium [40], tantalum [50], and molybdenum [42], which possess high melting points of over 1500°C or even over 3000°C, have been used for demonstrating high-temperature MPA for TPV applications. Using platinum as the constituent of MPAs, Woolf and his colleagues demonstrated an infrared high-temperature emitter with operation temperature over 1000°C [45]. As shown in Fig. 6.4c, the emitter consists of an array of platinum disks on a platinum-supported alumina spacer. The entire structure was encapsulated by an additional conformal layer of alumina with thickness of 150-nm. To characterize the thermal-to-electrical power conversion efficiency, the sample was mounted on a cartridge heater with a five-junction InGaAs PV cell placed about 22.35 mm above the emitter (Fig. 6.4b). A known Si emitter and a thermal power meter were used for calibration. Based on the measured short IV curves as a function of the SE temperature as shown in Fig. 6.4c, the corresponding thermal-to-electric conversion efficiency can be up to 24.1% at 1055°C. Figure 6.4d shows the emission performance of the TPV system.

In addition to the refractory metals, refractory plasmonic materials with plasmonic frequencies in the visible and near-infrared wavelength, such as TiN, aluminum zinc oxide (AZO), also show melting points over 2200°C [33, 48]. Similarly, the plasmonic nature and large intrinsic loss of these materials can also lead to large absorption spanning a broad band. Li and coworkers demonstrated a high-temperature broadband absorber can endure a laser illumination of 6.67 W/cm^2 after 800°C annealing of TiN, (Fig. 6.4f) while a gold based absorber was significantly degraded under the same process, with the gold structure peeling off from the sub-

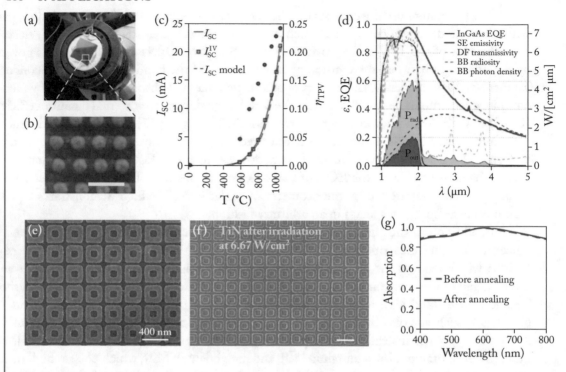

Figure 6.4: (a) Image of a selective emitter (SE) mounted on a cartridge heater. (b) SEM image of the metamaterial SE. The scale bar is 1 μm. (c) Short-circuit current as a function of the SE temperature. The right vertical axis shows the corresponding thermal-to-electric conversion efficiency η_{TPV} indicated as circles in the plot. (d) Spectral properties of the TPV system. SE: selective emitter DF: dielectric filter BB: black body. (e) SEM image of the fabricated TiN absorber with arrayed square ring-like structures. (f) SEM image of the TiN absorber shot by a 550-nm laser with an intensity of 6.67 W/cm². The scale bar is 400-nm. (g) Measured absorption before and after annealing at 800°C. Reprinted with permission from: (a)–(d) Ref. [45] © 2018 OSA; (e)–(g) Ref. [48] © 2014 Wiley-VCH.

strate [48]. As shown in Fig. 6.4e, the high-temperature MPAs consists of arrayed square ring-like TiN cells on a 60-nm silicon oxide spacer with a continuous 150-nm TiN thin film as the ground plane. Figure 6.4g shows the measured absorption before and after annealing at 800°C for 8 h with negligible change, while the optical performance of the gold sample is degraded, due to the reshaping of the structure after melting.

For practical TPV applications, there are at least five crucial requirements for the emitters: optical performance, ability to scale to large areas, log-term high-temperature stability, ease of integration, and low cost [51]. However, few of the demonstrated high-temperature MPA-based selective emitters satisfy all of these five criteria. The long-term operational stability of the artificial composites with various materials should be further investigated. Also, since the feature size of most of selective emitters working in the near-infrared or visible range is sub-micron scale, the ability to pattern in a large-area format has to be addressed before further commercialization.

6.3   IMAGING

Exploration of optical components with the ability for efficient information processing has long been a goal for scientists and engineers over the last several decades. Imaging can provide temporal and spectral information, and is one of the most important applications. Due to the spectrally selective nature of metamaterials, MPAs have great potential to form active and passive components for imaging from microwave frequencies to visible wavelengths.

As passive imaging devices, MPAs may operate as the key element of a microbolometer in which the incident EM radiation is efficiently absorbed and sensed by a thermometer. Such bolometer MMA elements may be fashioned as MEMS structures [52], phase change materials [53], and pyroelectric materials [54]. An initial conceptual demonstration showed that integration of an MPA onto a microbridge supported plate can significantly enhance the responsivity and spectral selectivity compared to other commercially available bolometers [55]. The measured results showed that the integration of an MPA with absorption peak of 57% at 2.5 THz achieves a minimum NEP of 37 pW/$\sqrt{\text{Hz}}$ and a thermal time constant of 68 ms. Since the fabrication process is compatible to the IC process, the monolithic detector can be easily scaled to a focal plane array through tessellating multiple elements together and adding standard circuit read-out, forming a terahertz imager. An alternative approach to detect EM radiation is to mix the absorption and sensing elements together, which additionally forms a more compact device. For example, the pyroelectric effect relies on a temperature gradient which induces a change in polarization, and broadband absorber coated bulk pyroelectric materials, like $LiNbO_3$ slabs, are commercially available and permit detection at room temperature from microwave to visible wavelengths. However, because of the large thermal mass which results from the relatively thick slab, the low responsivity of the sensing element requires a significantly large gain in the read-out, which could further reduce the signal-to-noise ratio. With the aid of sputtering or ion-slicing techniques, pyroelectric materials can be fabricated with sub-micron thickness, thus lowering the thermal mass. The thin film can be directly integrated into a MPA as the spacer [56]. With the judicious design of the metamaterial absorber, as shown in Fig. 6.5a–c, 82.5% of the total incident power can be absorbed by the pyroelectric spacer directly. In this way, the converted heat immediately produces the polarization without extra delay from the heat transferred from the top and bottom metal layer. In addition, the overlap between the DC field and the optical absorption region inside the spacer can further reduce the effective thermal radiative area since nearly 90% of the absorbed power in the $LiNO_3$ only occurs in 2.4% of the volume, thereby leading to a decrease in limiting noise equivalent power (NEP). The experimental measurements showed about 86% of absorption at 10.73 μm with a thermal time

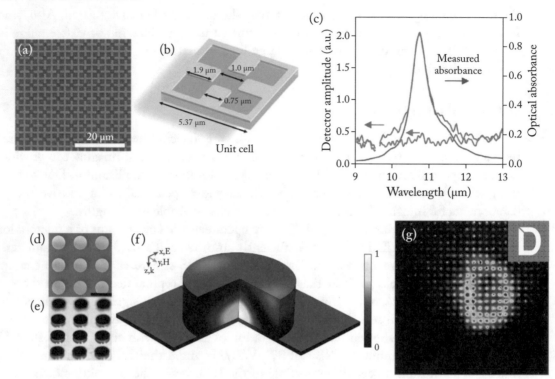

Figure 6.5: (a) An SEM image of the top layer of the MPA. (b) Schematic of the unit cell of the designed MPA with a 550-nm-thick LiNO₃ spacer layer. (c) Detector response (blue curve) of MPA detector matches well with the measured optical absorbance (red curve). The green curve shows the detector response of a non-resonant grid. (d) SEM image of the fabricated perfect absorber with array of silicon cylinders. (e) Oblique view of the fabricated sample. (f) Simulated power loss density inside the absorber. (g) False color THz image of a "D" letter using the THz to IR conversion technique. Reprinted with permission from (a)–(c) Ref. [56] © 2017 OSA, (d)–(g) Ref. [57] © 2017 OSA.

constant of 28.9 ms and a room temperature detectivity, D^*, close to 10^7 cm W/$\sqrt{\text{Hz}}$, which is approximately one order of magnitude larger than commercial detectors.

In most cases, energy absorbed with the metamaterial is dissipated as heat, which re-emits into space with a spectrum determined by Planck's radiation law. With the emission peaked in the long-wave infrared region, the signature can be directly captured by a thermal camera. The integration of an infrared camera has shown significant potential for imaging in spectral ranges which lack commercial availability of FPAs, such as millimeter and terahertz frequencies. Kuznetsov and his colleagues demonstrated a terahertz-to-infrared converter which absorbs incident terahertz waves and re-emit thermal radiation for imaging. To ensure efficient infrared emission from the heated sample, a thin layer of graphite was coated on the backside of the back metal-plane. Due to the narrowband resonance of the absorption, a multispectral THz FPA with 24×24 pixels has been demonstrated by inlaying three different absorber arrays operating at 0.3, 0.33, and 0.36 THz, respectively. Such an imaging scheme presents a new way for a flexible design and a low-cost implementation for terahertz imaging. However, the large thermal conductivity of the continuous back metal plane quickly diffuses the heat to neighbor pixels, thus leading to thermal spreading and lowering the imaging resolution. Alternatively, a perfect absorber fashioned entirely from sub-wavelength all-dielectric cylinders could significantly lower the thermal conductivity by three orders of magnitude compared to metal-based counterparts. Since the dielectric cylinders are isolated from each other, the channels for transferring converted heat from the absorbed energy are reduced to convection and thermal radiation. The reduced thermal conductance results in a higher temperature inside the cylinders for infrared camera detection, thus increasing the responsivity. Figure 6.5d shows an SEM image of the fabricated silicon cylinder array on an ultrathin PDMS substrate, where the cylinder height was 85 μm, diameter of 212 μm, and periodicity of 330 μ [57]. The experimental characterization of the fabricated absorber exhibited the peak absorption as high as 96% at the resonant frequency of 603 GHz and achieved a thermal responsivity of 2.16×10^4 K/W at a modulation of 1/4 Hz at room temperature and under standard atmospheric pressure. Figure 6.5g shows a false-color of a THz image with an object of a "D" letter with linewidth of 5 mm.

The rapid development of smartphone technology has led to modern cellular telephones possessing digital cameras with tens of megapixels, and has enabled the explosion of data in our daily lives. However, smartphone space is at a premium, which limits the available lateral area of the digital camera, thereby reducing available light flux and reducing the number of pixels. Alternatively, computational imaging may be used to overcome the limited light flux, and relatively low number of pixels. In contrast to traditional imaging, computational imaging involves the joint design of the physical sensing system and post-processing unit. Benefits from advances in algorithms and computing hardware, have enabled computational imaging system to provide significantly enhanced capabilities, but with reduced requirements in size, weight, power, and cost. Recently, computational imaging has garnered intensive attention for imaging in the infrared and terahertz regions, where large format cameras are scarce [59, 60]. Figure 6.6a illustrates a terahertz single-pixel imaging scheme using a metamaterial absorber based spatial light modulator (SLM). The spatial light modulator consists of 8×8 tunable metamaterial pixels with metamaterial absorbers in each pixel. The reflectance of each pixel can be independently controlled via a gating voltage from external electronics. In the experiment, a terahertz beam from an incoherent mercury-arc lamp was collimated to illuminate an object, the image of which was multiplexed by the programmed SLM. To obtain Hadamard masking with $[1, -1]$ on each pixel, a lock-in detection scheme with modulating in and out of phase with respect to a reference was implemented. A liquid-helium cooled bolometer was used as the single pixel to collect the multiplexed signal. Finally, as shown in Fig. 6.6e, high-fidelity terahertz images of the

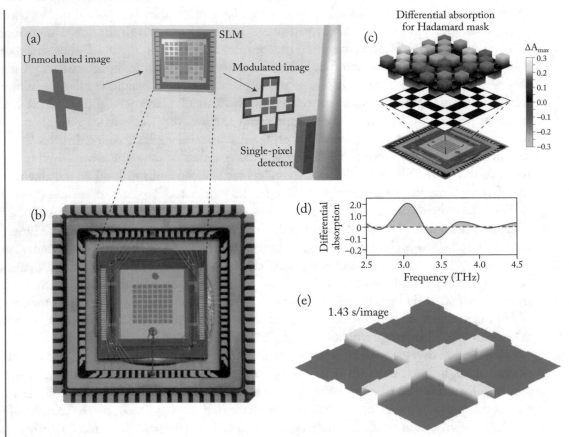

Figure 6.6: (a) A schematic of the single-pixel imaging process using a metamaterial based spatial light modulator coding the object image. (b) Photograph of the 8×8 SLM device. (c) Spatial map of maximum differential absorption for an example of Hadamard mask. (d) Measured differential absorption spectrum between 0 V and 15 V reverse biases. (e) Reconstructed image of an object with an inverse cross on a metal plate. The single-pixel imaging process involves 64 different Hadamard mask with a total image acquisition time of 1.43 s Ref. [58]. Reprinted with permission from © 2014 NPG.

object were retrieved—although illumination power was only 67 nW—using both Hadamard and compressive imaging mask encodings. With the same technique, the single-pixel imaging based on tunable metamaterial absorbers was also experimentally demonstrated in the infrared via gating the Fermi level of graphene to modulate the reflectance of metamaterial pixels. Further, metamaterial absorber based SLMs can be utilized for single-pixel parallel imaging using multiplexing techniques to decrease the image acquisition time. By changing the modulation phase on each pixel, the SLM can also be used to carry out phase-shift-key (PSK) single-pixel imaging, including quadrature-amplitude modulation (QAM), and 8-PSK modulation. On the other hand, different masks can be encoded with orthogonal modulation frequencies to modify the response of pixels. With this orthogonal frequency-division multiplexing (OFDM) method, several masks can be encoded on the SLM simultaneously, thus effectively increasing the image acquisition speed but at a cost of SNR.

6.4   CONCLUSION

Although significant progress has been made to design and use metamaterial perfect absorbers for various applications, it is still technically challenging to use a conventional design strategy to match the EQE of PV cells exactly, i.e., searching the best fit by sweeping the dimensions of metamaterial supercells in a very large geometrical space. Such an involving optimization process requires significant computation time and cost. Recent advances in the data-driven deep learning technology have opened a new paradigm to fashion metamaterials with high complexity and multi-constraints. With the aid of the deep learning technique, the relationships between the geometry at the input port and electromagnetic response at the output can be mimicked by an artificial neural network. Then the metamaterial geometry with the corresponding objective response can be inferred with speed several orders of magnitude faster using efficient inverse design strategies, including tandem neural network [61], generative adversarial network [62], neural-adjoint method [63], and adversarial autoencoder [64]. These data-driven design methodologies are a promising path to transform the way metamaterials are simulated and designed, and will facilitate metamaterial integration into practical applications.

6.5   REFERENCES

[1] Zhaoyi Li and Nanfang Yu. Modulation of mid-infrared light using graphene-metal plasmonic antennas. *Applied Physics Letters*, 102(13):131108, April 2013. DOI: 10.1063/1.4800931 157

[2] Kebin Fan, Jonathan Suen, Xueyuan Wu, and Willie J. Padilla. Graphene metamaterial modulator for free-space thermal radiation. *Optics Express*, 24(22):25189, October 2016. DOI: 10.1364/oe.24.025189 157

[3] Shicheng Fan and Yaoliang Song. Bandwidth-enhanced polarization-insensitive metamaterial absorber based on fractal structures. *Journal of Applied Physics*, 123(8):085110, February 2018. DOI: 10.1063/1.5004629 157

[4] Na Liu, Martin Mesch, Thomas Weiss, Mario Hentschel, and Harald Giessen. Infrared perfect absorber and its application as plasmonic sensor. *Nano Letters*, 10(7):2342–2348, July 2010. DOI: 10.1021/nl9041033 157, 158, 159, 163

[5] Khagendra Bhattarai, Zahyun Ku, Sinhara Silva, Jiyeon Jeon, Jun Oh Kim, Sang Jun Lee, Augustine Urbas, and Jiangfeng Zhou. A large-area, mushroom-capped plasmonic perfect

absorber: Refractive index sensing and Fabry-Perot cavity mechanism. *Advanced Optical Materials*, 3(12):1779–1786, December 2015. DOI: 10.1002/adom.201500231 158, 159

[6] Joshua A. Mason, Graham Allen, Viktor A. Podolskiy, and Daniel Wasserman. Strong coupling of molecular and mid-infrared perfect absorber resonances. *IEEE Photonics Technology Letters*, 24(1):31–33, January 2012. DOI: 10.1109/lpt.2011.2171942 158, 159

[7] Andreas Tittl, Ann-Katrin U. Michel, Martin Schäferling, Xinghui Yin, Behrad Gholipour, Long Cui, Matthias Wuttig, Thomas Taubner, Frank Neubrech, and Harald Giessen. A switchable mid-infrared plasmonic perfect absorber with multispectral thermal imaging capability. *Advanced Materials*, 27(31):4597–4603, August 2015. DOI: 10.1002/adma.201502023 157

[8] Fei Cheng, Xiaodong Yang, and Jie Gao. Enhancing intensity and refractive index sensing capability with infrared plasmonic perfect absorbers. *Optics Letters*, 39(11):3185, June 2014. DOI: 10.1364/ol.39.003185 159

[9] Longqing Cong, Siyu Tan, Riad Yahiaoui, Fengping Yan, Weili Zhang, and RanJan Singh. Experimental demonstration of ultrasensitive sensing with terahertz metamaterial absorbers: A comparison with the metasurfaces. *Applied Physics Letters*, 106(3):031107, January 2015. DOI: 10.1063/1.4906109 159

[10] Kai Chen, Thang Duy Dao, Satoshi Ishii, Masakazu Aono, and Tadaaki Nagao. Infrared aluminum metamaterial perfect absorbers for plasmon-enhanced infrared spectroscopy. *Advanced Functional Materials*, 25(42):6637–6643, November 2015. DOI: 10.1002/adfm.201501151 159

[11] Mark W. Knight, Heidar Sobhani, Peter Nordlander, and Naomi J. Halas. Photodetection with active optical antennas. *Science*, 332(6030):702–704, May 2011. DOI: 10.1126/science.1203056 160

[12] Keng Te Lin, Hsuen Li Chen, Yu Sheng Lai, and Chen Chieh Yu. Silicon-based broadband antenna for high responsivity and polarization-insensitive photodetection at telecommunication wavelengths. *Nature Communications*, 5(1):1–10, February 2014. DOI: 10.1038/ncomms4288 160, 161

[13] Wei Li and Jason Valentine. Metamaterial perfect absorber based hot electron photodetection. *Nano Letters*, 14(6):3510–3514, June 2014. DOI: 10.1021/nl501090w 160, 161

[14] Yuhua Lu, Wen Dong, Zhuo Chen, Anders Pors, Zhenlin Wang, and Sergey I. Bozhevolnyi. Gap-plasmon based broadband absorbers for enhanced hot-electron and photocurrent generation. *Scientific Reports*, 6(1):1–9, July 2016. DOI: 10.1038/srep30650 160, 161

[15] Long Wen, Yifu Chen, Li Liang, and Qin Chen. Hot electron harvesting via photoelectric ejection and photothermal heat relaxation in hotspots-enriched plasmonic/photonic disordered nanocomposites. *ACS Photonics*, 5(2):581–591, February 2018. DOI: 10.1021/acsphotonics.7b01156 160, 161, 162

[16] Charlene Ng, Jasper J. Cadusch, Svetlana Dligatch, Ann Roberts, Timothy J. Davis, Paul Mulvaney, and Daniel E. Gómez. Hot carrier extraction with plasmonic broadband absorbers. *ACS Nano*, 10(4):4704–4711, April 2016. DOI: 10.1021/acsnano.6b01108 160

[17] Matthew E. Sykes, Jon W. Stewart, Gleb M. Akselrod, Xiang Tian Kong, Zhiming Wang, David J. Gosztola, Alex B. F. Martinson, Daniel Rosenmann, Maiken H. Mikkelsen, Alexander O. Govorov, and Gary P. Wiederrecht. Enhanced generation and anisotropic Coulomb scattering of hot electrons in an ultra-broadband plasmonic nanopatch metasurface. *Nature Communications*, 8(1):1–11, December 2017. DOI: 10.1038/s41467-017-01069-3 162

[18] Hayk Harutyunyan, Alex B. F. Martinson, Daniel Rosenmann, Larousse Khosravi Khorashad, Lucas V. Besteiro, Alexander O. Govorov, and Gary P. Wiederrecht. Anomalous ultrafast dynamics of hot plasmonic electrons in nanostructures with hot spots. *Nature Nanotechnology*, 10(9):770–774, September 2015. DOI: 10.1038/nnano.2015.165 162

[19] Van-Duong Dao and Ho-Suk Choi. Carbon-based sunlight absorbers in solar-driven steam generation devices. *Global Challenges*, 2(2):1700094, February 2018. DOI: 10.1002/gch2.201700094 162

[20] Kyuyoung Bae, Gumin Kang, Suehyun K. Cho, Wounjhang Park, Kyoungsik Kim, and Willie J. Padilla. Flexible thin-film black gold membranes with ultrabroadband plasmonic nanofocusing for efficient solar vapour generation. *Nature Communications*, 6:10103, December 2015. DOI: 10.1038/ncomms10103 162, 164

[21] Lin Zhou, Yingling Tan, Jingyang Wang, Weichao Xu, Ye Yuan, Wenshan Cai, Shining Zhu, and Jia Zhu. 3D self-assembly of aluminium nanoparticles for plasmon-enhanced solar desalination. *Nature Photonics*, 10(6):393–398, April 2016. DOI: 10.1038/nphoton.2016.75 163, 164

[22] Satoshi Ishii, Ramu Pasupathi Sugavaneshwar, and Tadaaki Nagao. Titanium nitride nanoparticles as plasmonic solar heat transducers. *The Journal of Physical Chemistry C*, 120(4):2343–2348, February 2016. DOI: 10.1021/acs.jpcc.5b09604 162

[23] Muhammad Usman Farid, Jehad A. Kharraz, Peng Wang, and Alicia Kyoungjin An. High-efficiency solar-driven water desalination using a thermally isolated plasmonic membrane. *Journal of Cleaner Production*, 271:122684, October 2020. DOI: 10.1016/j.jclepro.2020.122684 162

[24] Lulu Zhang, Jun Xing, Xinglin Wen, Jianwei Chai, Shijie Wang, and Qihua Xiong. Plasmonic heating from indium nanoparticles on a floating microporous membrane for enhanced solar seawater desalination. *Nanoscale*, 9(35):12843–12849, September 2017. DOI: 10.1039/c7nr05149b 162

[25] Lin Zhou, Yingling Tan, Dengxin Ji, Bin Zhu, Pei Zhang, Jun Xu, Qiaoqiang Gan, Zongfu Yu, and Jia Zhu. Self-assembly of highly efficient, broadband plasmonic absorbers for solar steam generation. *Science Advances*, 2(4):e1501227, April 2016. DOI: 10.1126/sciadv.1501227 163

[26] Lin Zhou, Shendong Zhuang, Chengyu He, Yingling Tan, Zhenlin Wang, and Jia Zhu. Self-assembled spectrum selective plasmonic absorbers with tunable bandwidth for solar energy conversion. *Nano Energy*, 32:195–200, February 2017. DOI: 10.1016/j.nanoen.2016.12.031 163

[27] Xinzhi Wang, Yurong He, Xing Liu, Gong Cheng, and Jiaqi Zhu. Solar steam generation through bio-inspired interface heating of broadband-absorbing plasmonic membranes. *Applied Energy*, 195:414–425, June 2017. DOI: 10.1016/j.apenergy.2017.03.080 163

[28] Liangliang Zhu, Minmin Gao, Connor Kang Nuo Peh, Xiaoqiao Wang, and Ghim Wei Ho. Self-contained monolithic carbon sponges for solar-driven interfacial water evaporation distillation and electricity generation. *Advanced Energy Materials*, 8(16), June 2018. DOI: 10.1002/aenm.201702149 163

[29] Mingwei Zhu, Yiju Li, Fengjuan Chen, Xueyi Zhu, Jiaqi Dai, Yongfeng Li, Zhi Yang, Xuejun Yan, Jianwei Song, Yanbin Wang, Emily Hitz, Wei Luo, Minhui Lu, Bao Yang, and Liangbing Hu. Plasmonic wood for high-efficiency solar steam generation. *Advanced Energy Materials*, 8(4), February 2018. DOI: 10.1002/aenm.201701028

[30] Yang Yang, Ruiqi Zhao, Tengfei Zhang, Kai Zhao, Peishuang Xiao, Yanfeng Ma, Pulickel M. Ajayan, Gaoquan Shi, and Yongsheng Chen. Graphene-based standalone solar energy converter for water desalination and purification. *ACS Nano*, 12(1):829–835, January 2018. DOI: 10.1021/acsnano.7b08196 163

[31] Chengmin Sheng, Ning Yang, Yutao Yan, Xiaoping Shen, Chunde Jin, Zhe Wang, and Qingfeng Sun. Bamboo decorated with plasmonic nanoparticles for efficient solar steam generation. *Applied Thermal Engineering*, 167:114712, February 2020. DOI: 10.1016/j.applthermaleng.2019.114712 163

[32] Weiren Zhu, Ivan D. Rukhlenko, Fajun Xiao, Chong He, Junping Geng, Xianling Liang, Malin Premaratne, and Ronghong Jin. Multiband coherent perfect absorption in a water-based metasurface. *Optics Express*, 25(14):15737, July 2017. DOI: 10.1364/oe.25.015737 163

[33] Sean Molesky, Christopher J. Dewalt, and Zubin Jacob. High temperature epsilon-near-zero and epsilon-near-pole metamaterial emitters for thermophotovoltaics. *Optics Express*, 21(S1):A96, January 2013. DOI: 10.1364/oe.21.000a96 163, 165

[34] Zhi-Yuan Li. Modified thermal radiation in three-dimensional photonic crystals. *Physical Review B*, 66:241103, December 2002. DOI: 10.1103/physrevb.66.241103 163

[35] Pei-En Chang, Yu-Wei Jiang, Hung-Hsin Chen, Yi-Tsung Chang, Yi-Ting Wu, Lawrence Dah-Ching Tzuang, Yi-Han Ye, and Si-Chen Lee. Wavelength selective plasmonic thermal emitter by polarization utilizing fabry-pérot type resonances. *Applied Physics Letters*, 98(7):073111, 2011. DOI: 10.1063/1.3537807 163

[36] Xianliang Liu, Tatiana Starr, Anthony F. Starr, and Willie J. Padilla. Infrared spatial and frequency selective metamaterial with near-unity absorbance. *Physical Review Letters*, 104(20):207403, May 2010. DOI: 10.1103/physrevlett.104.207403 163

[37] Carl Hägglund and S. Peter Apell. Plasmonic near-field absorbers for ultrathin solar cells. *The Journal of Physical Chemistry Letters*, 3(10):1275–1285, May 2012. DOI: 10.1021/jz300290d 163

[38] Qiuqun Liang, Taisheng Wang, Zhenwu Lu, Qiang Sun, Yongqi Fu, and Weixing Yu. Metamaterial-based two dimensional plasmonic subwavelength structures offer the broadest waveband light harvesting. *Advanced Optical Materials*, 1(1):43–49, January 2013. DOI: 10.1002/adom.201200009 163

[39] Corey Shemelya, Dante DeMeo, Nicole Pfiester Latham, Xueyuan Wu, Chris Bingham, Willie Padilla, and Thomas E. Vandervelde. Stable high temperature metamaterial emitters for thermophotovoltaic applications. *Applied Physics Letters*, 104(20):201113, May 2014. DOI: 10.1063/1.4878849 163, 165

[40] Hao Wang, Vijay Prasad Sivan, Arnan Mitchell, Gary Rosengarten, Patrick Phelan, and Liping Wang. Highly efficient selective metamaterial absorber for high-temperature solar thermal energy harvesting. *Solar Energy Materials and Solar Cells*, 137:235–242, June 2015. DOI: 10.1016/j.solmat.2015.02.019 163, 165

[41] Ankit Vora, Jephias Gwamuri, Nezih Pala, Anand Kulkarni, Joshua M. Pearce, and Durdu Ö. Güney. Exchanging ohmic losses in metamaterial absorbers with useful optical absorption for photovoltaics. *Scientific Reports*, 4(1):4901, May 2015. DOI: 10.1038/srep04901 163

[42] Takahiro Yokoyama, Thang Duy Dao, Kai Chen, Satoshi Ishii, Ramu Pasupathi Sugavaneshwar, Masahiro Kitajima, and Tadaaki Nagao. Spectrally selective mid-infrared thermal emission from molybdenum plasmonic metamaterial operated up

to 1000°C. *Advanced Optical Materials*, 4(12):1987–1992, December 2016. DOI: 10.1002/adom.201600455 163, 165

[43] P. N. Dyachenko, S. Molesky, A. Yu Petrov, M. Störmer, T. Krekeler, S. Lang, M. Ritter, Z. Jacob, and M. Eich. Controlling thermal emission with refractory epsilon-near-zero metamaterials via topological transitions. *Nature Communications*, 7(1):11809, December 2016. DOI: 10.1038/ncomms11809 163, 165

[44] Cheng Shi, Nathan H. Mahlmeister, Isaac J. Luxmoore, and Geoffrey R. Nash. Metamaterial-based graphene thermal emitter. *Nano Research*, 11(7):3567–3573, July 2018. DOI: 10.1007/s12274-017-1922-7 163

[45] David N. Woolf, Emil A. Kadlec, Don Bethke, Albert D. Grine, John J. Nogan, Jeffrey G. Cederberg, D. Bruce Burckel, Ting Shan Luk, Eric A. Shaner, and Joel M. Hensley. High-efficiency thermophotovoltaic energy conversion enabled by a metamaterial selective emitter. *Optica*, 5(2):213, February 2018. DOI: 10.1364/optica.5.000213 163, 165, 166

[46] Chun-Chieh Chang, Wilton J. M. Kort-Kamp, John Nogan, Ting S. Luk, Abul K. Azad, Antoinette J. Taylor, Diego A. R. Dalvit, Milan Sykora, and Hou-Tong Chen. High-temperature refractory metasurfaces for solar thermophotovoltaic energy harvesting. *Nano Letters*, 18(12):7665–7673, December 2018. DOI: 10.1021/acs.nanolett.8b03322 163, 165

[47] Xianliang Liu, Talmage Tyler, Tatiana Starr, Anthony F. Starr, Nan Marie Jokerst, and Willie J. Padilla. Taming the blackbody with infrared metamaterials as selective thermal emitters. *Physical Review Letters*, 107(4):045901, July 2011. DOI: 10.1103/physrevlett.107.045901 165

[48] Wei Li, Urcan Guler, Nathaniel Kinsey, Gururaj V. Naik, Alexandra Boltasseva, Jianguo Guan, Vladimir M. Shalaev, and Alexander V. Kildishev. Refractory plasmonics with titanium nitride: Broadband metamaterial absorber. *Advanced Materials*, 26(47):7959–7965, December 2014. DOI: 10.1002/adma.201401874 165, 166

[49] Chihhui Wu, Burton Neuner III, Jeremy John, Andrew Milder, Byron Zollars, Steve Savoy, and Gennady Shvets. Metamaterial-based integrated plasmonic absorber/emitter for solar thermo-photovoltaic systems. *Journal of Optics*, 14(2):024005, February 2012. DOI: 10.1088/2040-8978/14/2/024005 165

[50] Yang Li, Dezhao Li, Dan Zhou, Cheng Chi, Shihe Yang, and Baoling Huang. Efficient, scalable, and high-temperature selective solar absorbers based on hybrid-strategy plasmonic metamaterials. *Solar RRL*, 2(8):1800057, August 2018. DOI: 10.1002/solr.201800057 165

[51] Reyu Sakakibara, Veronika Stelmakh, Walker R. Chan, Michael Ghebrebrhan, John D. Joannopoulos, Marin Soljačić, and Ivan Čelanović. Practical emitters for thermophotovoltaics: A review. *Journal of Photonics for Energy*, 9(03):1, February 2019. DOI: 10.1117/1.jpe.9.032713 166

[52] Frank Niklaus, Christian Vieider, and Henrik Jakobsen. MEMS-based uncooled infrared bolometer arrays: A review. In Jung-Chih Chiao, Xuyuan Chen, Zhaoying Zhou, and Xinxin Li, editors, *MEMS/MOEMS Technologies and Applications III*, volume 6836, page 68360D, SPIE, November 2007. DOI: 10.1117/12.755128 167

[53] Sihai Chen, Hong Ma, Xinjian Yi, Tao Xiong, Hongcheng Wang, and Caijun Ke. Smart VO2 thin film for protection of sensitive infrared detectors from strong laser radiation. *Sensors and Actuators A: Physical*, 115(1):28–31, 2004. DOI: 10.1016/j.sna.2004.03.018 167

[54] D. L. Polla, Chian-ping Ye, and Takashi Tamagawa. Surface-micromachined pb-tio3 pyroelectric detectors. *Applied Physics Letters*, 59(27):3539–3541, 1991. DOI: 10.1063/1.105650 167

[55] Thomas Maier and Hubert Brückl. Wavelength-tunable microbolometers with metamaterial absorbers. *Optics Letters*, 34(19):3012, October 2009. DOI: 10.1364/ol.34.003012 167

[56] J. Y. Suen, K. Fan, J. Montoya, C. Bingham, Vincent Stenger, S. Sriram, and W. J. Padilla. Multifunctional metamaterial pyroelectric infrared detectors. *Optica*, 4(2):276, February 2017. DOI: 10.1364/optica.4.000276 167, 168

[57] Kebin Fan, Jonathan Y. Suen, Xinyu Liu, and Willie J. Padilla. All-dielectric metasurface absorbers for uncooled terahertz imaging. *Optica*, 4(6):601, June 2017. DOI: 10.1364/optica.4.000601 168, 169

[58] Claire M. Watts, David Shrekenhamer, John Montoya, Guy Lipworth, John Hunt, Timothy Sleasman, Sanjay Krishna, David R. Smith, and Willie J. Padilla. Terahertz compressive imaging with metamaterial spatial light modulators. *Nature Photonics*, 8(8):605–609, June 2014. DOI: 10.1038/nphoton.2014.139 170

[59] A. Rogalski. Recent progress in infrared detector technologies. *Infrared Physics and Technology*, 54(3):136–154, May 2011. DOI: 10.1016/j.infrared.2010.12.003 169

[60] David Shrekenhamer, Claire M. Watts, and Willie J. Padilla. Terahertz single pixel imaging with an optically controlled dynamic spatial light modulator. *Optics Express*, 21(10):12507–18, May 2013. DOI: 10.1364/oe.21.012507 169

[61] Zhaocheng Liu, Dayu Zhu, Sean P. Rodrigues, Kyu-Tae Lee, and Wenshan Cai. Generative model for the inverse design of metasurfaces. *Nano Letters*, 18(10):6570–6576, October 2018. DOI: 10.1021/acs.nanolett.8b03171 171

[62] Jiaqi Jiang, David Sell, Stephan Hoyer, Jason Hickey, Jianji Yang, and Jonathan A. Fan. Free-form diffractive metagrating design based on generative adversarial networks. *ACS Nano*, 13(8):8872–8878, August 2019. DOI: 10.1021/acsnano.9b02371 171

[63] Simiao Ren, Willie Padilla, and Jordan Malof. Benchmarking deep inverse models over time, and the neural-adjoint method. In H. Larochelle, M. Ranzato, R. Hadsell, M. F. Balcan, and H. Lin, editors, *Advances in Neural Information Processing Systems*, volume 33, pages 38–48, Curran Associates, Inc., 2020. 171

[64] Zhaxylyk A. Kudyshev, Alexander V. Kildishev, Vladimir M. Shalaev, and Alexandra Boltasseva. Machine-learning-assisted metasurface design for high-efficiency thermal emitter optimization. *Applied Physics Reviews*, 7(2):021407, June 2020. DOI: 10.1063/1.5134792 171

APPENDIX A

Temporal-Coupled Mode Theory

Amplitudes of a resonating system can be described by the temporal-coupled mode theory (TCMT). Here, we basically follow the description shown in Ref. [1] with a complex mode amplitude a associated to l ports:

$$\frac{da}{dt} = (-i\mathbf{\Omega} - \mathbf{\Gamma})\, a + \mathbf{K}^T s_{\text{in}} \tag{A.1}$$

$$s_{\text{out}} = \mathbf{C} s_{\text{in}} + \mathbf{D} a, \tag{A.2}$$

where matrices $\mathbf{\Omega}$ and $\mathbf{\Gamma}$ are the resonant frequencies and decay rates of the modes. Matrices $\mathbf{K}$ and $\mathbf{D}$ correspond to input and output coupling of resonant modes to the ports, respectively. s_{in} is a vector with elements representing inputs from each port where $|s_{\text{in}}|^2$ is equal to the input power, with a similar terms for the outputs s_{out}. Matrix $\mathbf{C}$ represents a direct non-resonant pathway from port to port. In general, for a lossless and reciprocal system, the matrix $\mathbf{C}$ is considered as unitary and symmetric. And in such a system, the energy of modes only decays as radiation and received by ports. Therefore, energy conservation and time-reversal symmetry still hold. And following relations can be derived as [1]

$$\mathbf{D}^\dagger \mathbf{D} = 2\mathbf{\Gamma}_r \tag{A.3a}$$
$$\mathbf{D} = \mathbf{K} \tag{A.3b}$$
$$\mathbf{C}\mathbf{D}^* = -\mathbf{D}, \tag{A.3c}$$

where $\mathbf{\Gamma}_r$ is the decay rates due to radiation into the ports. And these relations are still valid for a lossy resonator system when considering the internal loss rate $\mathbf{\Gamma}_i$ as a separate decay channel which couples to continuum states [1, 2]. From the TCMT equations, we may define a scattering matrix $\mathbf{S}$ as

$$\mathbf{S} \equiv \frac{s_{\text{out}}}{s_{\text{in}}} = \mathbf{C} + \mathbf{D}[-i\omega\mathbf{I} + i\mathbf{\Omega} + \mathbf{\Gamma}]^{-1}\mathbf{K}^T. \tag{A.4}$$

Considering a simple nonreciprocal resonating cavity with two ports symmetric with respect to its midplane, without loss of generality, both modes can be either even or odd. The parameters in the scattering matrix (Eq. (A.4)) are

$$\mathbf{\Omega} = \begin{bmatrix} \omega_1 & 0 \\ 0 & \omega_2 \end{bmatrix} \tag{A.5}$$

$$\Gamma = \Gamma_r + \Gamma_i = \begin{bmatrix} \gamma_1 + \delta_1 & \gamma_0 \\ \gamma_0^* & \gamma_2 + \delta_2 \end{bmatrix} \tag{A.6}$$

$$C = \begin{bmatrix} r & t \\ t & r \end{bmatrix} \tag{A.7}$$

$$D = \begin{bmatrix} D_{11} & D_{12} \\ D_{21} & D_{22} \end{bmatrix} = \begin{bmatrix} d_{11}e^{-i\theta_{11}} & d_{12}e^{-i\theta_{12}} \\ d_{21}e^{-i\theta_{21}} & d_{22}e^{-i\theta_{22}} \end{bmatrix}, \tag{A.8}$$

where ω_1 and ω_2 are the two resonance frequencies and δ_1 and δ_2 are the intrinsic material loss rates. γ_1 and γ_2 correspond to the modes decay rates to the ports, γ_0 defines the coupling between the two modes. The nonresonant complex reflection and transmission coefficients are denoted by r and t, respectively. For the coupling matrix D, θ_{ij} is the phase angle of matrix element D_{ij}, and d_{ij} is the coupling amplitude. The first number i in the subscripts indicates the port number and the second number j is associated to the mode. According to Eqs. (A.3)a and (A.3)c, we have

$$\begin{bmatrix} D_{11}^* & D_{21}^* \\ D_{12}^* & D_{22}^* \end{bmatrix} \begin{bmatrix} D_{11} & D_{12} \\ D_{21} & D_{22} \end{bmatrix} = 2 \begin{bmatrix} \gamma_1 & \gamma_0 \\ \gamma_0^* & \gamma_2 \end{bmatrix} \tag{A.9a}$$

$$\begin{bmatrix} r & t \\ t & r \end{bmatrix} \begin{bmatrix} D_{11}^* & D_{12}^* \\ D_{21}^* & D_{22}^* \end{bmatrix} = - \begin{bmatrix} D_{11} & D_{12} \\ D_{21} & D_{22} \end{bmatrix}. \tag{A.9b}$$

These two equations can be broken down into following equations as

$$d_{11}^2 + d_{21}^2 = 2\gamma_1 \tag{A.10a}$$

$$d_{12}^2 + d_{22}^2 = 2\gamma_2 \tag{A.10b}$$

$$d_{11}d_{12}e^{i(\theta_{11}-\theta_{12})} + d_{21}d_{22}e^{i(\theta_{21}-\theta_{22})} = 2\gamma_0 \tag{A.10c}$$

$$rD_{11}^* + tD_{21}^* = -D_{11} \tag{A.10d}$$

$$rD_{12}^* + tD_{22}^* = -D_{12} \tag{A.10e}$$

$$tD_{11}^* + rD_{21}^* = -D_{21} \tag{A.10f}$$

$$tD_{12}^* + rD_{22}^* = -D_{22}. \tag{A.10g}$$

Given the assumption of a mirror symmetry of the system, each mode would have the same coupling amplitude to the ports except the phase depending on the mode symmetry, i.e.,

$$d_{1i} = d_{2i} = \sqrt{\gamma_i} \tag{A.11a}$$

$$e^{i(\theta_{1i}-\theta_{2i})} = \pm 1 \tag{A.11b}$$

$$e^{i(\theta_{11}-\theta_{12})} + e^{i(\theta_{21}-\theta_{22})} = \frac{2\gamma_0}{\sqrt{\gamma_1\gamma_2}}, \tag{A.11c}$$

where $i = 1, 2$, and the plus sign in Eq. (A.11)b is for the even mode and the minus sign is for the odd mode. Then the coupling matrix D can be reorganized as

$$D = \begin{bmatrix} \sqrt{\gamma_1}e^{-i\theta_{11}} & \sqrt{\gamma_2}e^{-i(\theta_{22}-\beta_2)} \\ \sqrt{\gamma_1}e^{-i(\theta_{11}-\beta_1)} & \sqrt{\gamma_2}e^{-i\theta_{22}} \end{bmatrix}, \tag{A.12}$$

where β_j $(j = 1, 2)$ is the phase determined by the symmetry of jth mode. For example, for an even (odd) mode, $\beta_j = 0$ (π), which means the mode property symmetric (anti-symmetric) with respect to the ports. Based on Eqs. (A.10)d and (A.10)g, we can get the relations: $e^{-i2\theta_{11}} = -(r + te^{-i\beta_1})$ and $e^{-i2\theta_{22}} = -(r + te^{-i\beta_2})$. With these relations, we can calculate the scattering matrix as

$$\mathbf{S} = \mathbf{C} + \mathbf{D}\mathbf{M}^{-1}\mathbf{D}^{\mathrm{T}}, \tag{A.13}$$

where

$$\mathbf{M} = \begin{bmatrix} L_1 & \gamma_0 \\ \gamma_0^* & L_2 \end{bmatrix}$$
$$L_1 = -i(\omega - \omega_1) + \delta_1 + \gamma_1$$
$$L_2 = -i(\omega - \omega_2) + \delta_2 + \gamma_2.$$

Then

$$\mathbf{D}\mathbf{M}^{-1}\mathbf{D}^{\mathrm{T}} = -\mathbf{D}\mathbf{M}^{-1}\mathbf{D}^{\dagger}\mathbf{C}^{\mathrm{T}}$$
$$= \frac{1}{\det(\mathbf{M})} \begin{bmatrix} X_{11} & X_{12} \\ X_{21} & X_{22} \end{bmatrix} \begin{bmatrix} r & t \\ t & r \end{bmatrix}, \tag{A.15}$$

where

$$X_{11} = X_{22} = L_2\gamma_1 + L_1\gamma_2 - \gamma_1\gamma_2(1 + e^{i\beta})$$
$$= P_2\gamma_1 + P_1\gamma_2$$
$$X_{12} = X_{21} = L_2\gamma_1 e^{i\beta_1} + L_1\gamma_2 e^{i\beta_2} - \gamma_1\gamma_2(e^{\beta_1} + e^{-i\beta_2})$$
$$= P_2\gamma_1 e^{i\beta_1} + P_1\gamma_2 e^{i\beta_2},$$

where

$$P_1 = -i(\omega - \omega_1) + \delta_1 + \frac{1 - e^{i\beta}}{2}\gamma_1$$
$$P_2 = -i(\omega - \omega_2) + \delta_2 + \frac{1 - e^{i\beta}}{2}\gamma_2 \tag{A.17}$$
$$\beta = \beta_2 + \beta_1.$$

Finally, the scattering matrix is derived as

$$\mathbf{S} = \begin{bmatrix} r & t \\ t & r \end{bmatrix} - \frac{1}{\det(\mathbf{M})}\left\{ P_2\gamma_1 \begin{bmatrix} 1 & e^{i\beta_1} \\ e^{i\beta_1} & 1 \end{bmatrix} + P_1\gamma_2 \begin{bmatrix} 1 & e^{i\beta_2} \\ e^{i\beta_2} & 1 \end{bmatrix} \right\} \begin{bmatrix} r & t \\ t & r \end{bmatrix}. \tag{A.18}$$

A.1   REFERENCES

[1] W. Suh, Z. Wang, and S. Fan. Temporal coupled-mode theory and the presence of non-orthogonal modes in lossless multimode cavities. *IEEE Journal of Quantum Electronics*, 40(10):1511–1518, October 2004. DOI: 10.1109/jqe.2004.834773 179

[2] Sander A. Mann, Dimitrios L. Sounas, and Andrea Alù. Nonreciprocal cavities and the time-bandwidth limit. *Optica*, 6(1):104, January 2019. DOI: 10.1364/optica.6.000104 179

Authors' Biographies

WILLIE J. PADILLA

Willie J. Padilla is a Full Professor in the Department of ECE at Duke University with Physics M.S. and Ph.D. degrees from the University of California San Diego. He was a Director's Post-doctoral Fellow at Los Alamos National Laboratory. In 2007 he received a Young Investigator Award from the Office of Naval Research, and Presidential Early Career Award for Scientists and Engineers in 2011. In 2012 he was elected a Fellow of the Optical Society of America, and Kavli Frontiers of Science Fellow in 2013. Dr. Padilla was elevated to Senior Member of the SPIE in 2018, and is a Fellow of the American Physical Society. Professor Padilla is a Web of Science Highly Cited Researcher in the field of Physics in 2018 and 2019, has more than 200 peer-reviewed journal articles, 2 book chapters, and 7 issued patents. He heads a group working in the area of artificially structured systems including metamaterials with a focus on machine learning, computational imaging, spectroscopy, and energy.

KEBIN FAN

Kebin Fan is an Associate Professor in the School of Electronic Science and Engineering at Nanjing University, China. He received a Ph.D. degree in mechanical engineering from Boston University in 2012. He was a Post-Doctoral Researcher and then a Research Scientist with the Boston College and Duke University from 2012–2018. He was a Research Assistant Professor in the Department of ECE at Duke University before joining Nanjing University in 2020. He has published more than 50 peer-reviewed papers and 3 issued patents. His research interests include terahertz and infrared metamaterial devices, energy harvesting, imaging, and micro-/nanofabrication techniques.

Printed in the United States
by Baker & Taylor Publisher Services